Association for Women in Mathematics Series

Volume 12

Series Editor
Kristin Lauter
Microsoft Research
Redmond, Washington, USA

Association for Women in Mathematics Series

Focusing on the groundbreaking work of women in mathematics past, present, and future, Springer's Association for Women in Mathematics Series presents the latest research and proceedings of conferences worldwide organized by the Association for Women in Mathematics (AWM). All works are peer-reviewed to meet the highest standards of scientific literature, while presenting topics at the cutting edge of pure and applied mathematics. Since its inception in 1971, The Association for Women in Mathematics has been a non-profit organization designed to help encourage women and girls to study and pursue active careers in mathematics and the mathematical sciences and to promote equal opportunity and equal treatment of women and girls in the mathematical sciences. Currently, the organization represents more than 3000 members and 200 institutions constituting a broad spectrum of the mathematical community, in the United States and around the world.

More information about this series at http://www.springer.com/series/13764

Asli Genctav • Kathryn Leonard • Sibel Tari
Evelyne Hubert • Geraldine Morin
Noha El-Zehiry • Erin Chambers
Editors

Research in Shape Analysis

WiSH2, Sirince, Turkey, June 2016

Editors

Asli Genctav
Department of Computer Engineering
Middle East Technical University
Ankara, Turkey

Sibel Tari
Department of Computer Engineering
Middle East Technical University
Ankara, Turkey

Geraldine Morin
Toulouse Institute of Computer Science R
Toulouse, Garonne (Haute), France

Erin Chambers
Department of Computer Science
St Louis University
St Louis, Missouri, USA

Kathryn Leonard
Occidental College
Department of Computer Science
Los Angeles, CA, USA

Evelyne Hubert
INRIA Méditerranée
Sophia Antipolis, France

Noha El-Zehiry
Siemens Healthcare
Princeton, NJ, USA

ISSN 2364-5733   ISSN 2364-5741   (electronic)
Association for Women in Mathematics Series
ISBN 978-3-030-08360-1   ISBN 978-3-319-77066-6   (eBook)
https://doi.org/10.1007/978-3-319-77066-6

Preface

The second Women in Shape (WiSh) workshop was held at Nesin Math Village in Sirince, Turkey, on June 6–12, 2016. Twenty-two women researchers at different stages of their careers attended the WiSh-2 workshop, which was organized by the editors of this volume. The participants worked on one of the following projects.

- Convolution Skeletons for Shape Modeling —
  Team Leaders: Evelyne Hubert and Géraldine Morin
- Shape Complexity — Team Leaders: Erin Chambers and Kathryn Leonard
- Combinatorial Optimization for Shape Based Segmentation —
  Team Leader: Noha El-Zehiry
- Shape Optimization — Team Leader: Sibel Tari

There was also a junior group composed of graduate and undergraduate students who worked on the Mesh Saliency project defined by Sibel Tari.

The WiSh-2 workshop ended up as a great experience for widening the WiSh network as new collaborations were started and the existing ones were strengthened. The participants at early stages of their careers had the opportunity to work on the projects led by the senior researchers. This volume is composed of the research and survey papers on shape analysis and modeling, authored by the members of the WiSh network, primarily the research groups formed at the WiSh-2 workshop.

Nesin Math Village was a pleasant place for the workshop with its beautiful natural surroundings full of olive trees and its position with an ideal distance to historical (Ephesus ruins) and natural (Dilek Peninsula National Park) places. Special thanks are due to Asli Can Korkmaz and Ceren Aydin, who kindly considered all of our requests and provided full support from beginning to end.

We would like to thank the following organizations for their generous financial support of the workshop: The Scientific and Technological Research Council of Turkey (TUBITAK) via grant 114E204, Nesin Math Village, National Science Foundation, and the Association for Women in Mathematics.

Ankara, Turkey
October 2017

Asli Genctav

Workshop Participants and Affiliations at the Time of the Workshop

Rüyam Acar, Okan University, Turkey
Venera Adanova, Middle East Technical University, Turkey
Gulce Bal, Middle East Technical University, Turkey
Erin Chambers, Saint Louis University, USA
Ilke Demir, Purdue University, USA
Carlotta Domeniconi, George Mason University, USA
Noha El-Zehiry, Siemens Healthineers, USA
Tegan Emerson, Colorado State University, USA
Asli Genctav, Middle East Technical University, Turkey
Cindy Grimm, Oregon State University, USA
Evelyne Hubert, INRIA Méditerranée, France
Kathryn Leonard, Occidental College, USA
Géraldine Morin, IRIT - University of Toulouse, France
Monika Muszkieta, Wroclaw University of Science and Technology, Poland
Athina Panotopoulou, Dartmouth College, USA
Elissa Ross, MESH Consultants Inc., Canada
Irmak Saglam, Middle East Technical University, Turkey
Fadime Sener, University of Bonn, Germany
Sibel Tari, Middle East Technical University, Turkey
Kathrin Welker, Trier University, Germany
Ilkyaz Yasal, Middle East Technical University, Turkey
Bengisu Yılmaz, Sabanci University, Turkey

Introduction

Computer vision is an interdisciplinary field that aims to mimic the human visual perception and its ability to process and analyze objects from different perspectives. For example, object recognition is one of the interesting problems in computer vision; it refers to the ability of the computer to identify an object in an image or a video sequence. Such task can seamlessly be accomplished by the human visual cortex; however, it is still an open problem in the computer vision society. Humans can distinguish objects based on their physical properties such as shape, texture, and color. Therefore, a lot of research effort in the object recognition literature is dedicated to understanding these features. Shape is the characteristic to which we dedicate this book.

Shape analysis is one of the most challenging, yet most important, problems in image processing and computer vision. Object shape is invariant to similarity transformation; in other words, characterizing the shape of the object makes it possible to identify the object even if it is translated, rotated, or dilated, a property referred to as shape constancy.

These proceedings discuss a broad range of topics in shape analysis: Shape representation, shape complexity, and characterization in solving image processing problems such as image segmentation and registration. Furthermore, some practical applications of shape analysis will be exposed.

Shape representation is usually the first step in any shape analysis workflow. Any inaccuracies in representing the object shape would jeopardize the accuracy of the whole workflow. Therefore, two-thirds of these proceedings is dedicated to exploring shape representation (Chapters 1–4) and to understanding the shape complexity (Chapters 3–6). Then, applications such as image segmentation, registration (Chapters 7–8), and image deblurring (Chapter 9) are discussed. Last but not least, a survey that discusses shape patterns in digital fabrication is presented.

One of the simplest and most common shape representation models is skeleton representation. In Chapter 1, Chambers et al. examine the usefulness of blum medial axis in representing natural images. They also show how to combine geometric cues and aggregated medial fragments to obtain an integrated representation. Building upon skeletal representation, in Chapter 2, Panotopoulou et al. present an approach to construct a coarse quadrilateral mesh around the one-dimensional skeleton. The authors also show that their construction is optimal in the sense that there is no coarser mesh that can be obtained from the same skeletal representation.

Scaling up from two-dimensional space to three-dimensional space poses a higher level of complexity. Instead of analyzing points and skeletons, it becomes necessary to analyze surfaces and meshes. In Chapter 3, Surárez and Hubert investigate the convolutional surface shape representation in 3D. They advocate that any space of curves can be approximated by circular splines in a $\mathcal{G}^1$ fashion. Hence, an appealing convolution surface can be obtained using a lower number of basic skeleton elements yielding a better visual quality at a lower computational cost.

Once a shape representation is obtained, it is important to explore the complexity of the shape and identify the critical or highly complex regions to be handled properly. For example, if we are trying to perform shape matching using mesh representation of shapes, it might be beneficial to identify the complex areas (e.g., higher curvature areas) and have a denser mesh representation in these areas. Shape complexity is discussed in Chapters 4–6.

In Chapter 4, Chambers et al. explore shape complexity in two-dimensional representations. The authors examine skeleton-based and symmetry-based complexity measures, as well as measures derived from the boundary sampling. They apply these shape complexity measures to a library of shapes and investigate which shape aspects are captured by such complexity representations. Furthermore, they present a novel complexity measure based on the blum medial axis.

Shape topology is one of the important complexity measures. Chapters 5 and 6 analyze different aspects of shape topology. In Chapter 5, Acar and Sağırlı propose a filter-based approach to determine topologically critical regions. Defining such regions is of high importance in constraining phase field evolution. In computer graphics literature, shape topology is oftentimes used to match 3D meshes. One of the caveats associated with topological matching is that the geometry or the topology can be corrupted with noise. Robustness to noise is crucial to be able to match points that are semantically equivalent. In Chapter 6, Genctav and Tari introduce the concept of barrier structure to equate shape topologies in the existence of noise.

Understanding the object shape and its level of complexity can be useful in many applications including, but not limited to, object recognition, image segmentation, image registration, and image deblurring. Chapters 7–10 of these proceedings discuss various applications of shape analysis.

For example, in Chapter 7, Atta-Fosu and Guo explore the shape application in simultaneous segmentation and registration. Detection of the shape changes can help identify objects or understand the effect of particular transformations on a given image or object. Simultaneous segmentation and registration studies this problem. Segmentation refers to the delineation of the boundaries, which basically characterizes the shape of the object of interest. Registration refers to the process of aligning two images; typically the two images are referred to as reference image and target image. The registration process aligns images by applying transformations on the reference image; such transformations can be rigid global transformation such as rotation and translation and can also be small local deformations that cannot be captured by global transformations. In Chapter 7, the authors explain in a very elegant framework how to perform simultaneous segmentation and registration using fast Fourier transform and total variation. In their formulation, both tasks can be performed by minimizing a carefully designed energy function that consists of a data fidelity term and a regularization term. If the shape of the object is identified in the reference image, the segmented regions in the reference image can then be used as the data fidelity term to fit the segmentation contours in the target image, then a nonlinear elasticity regularization penalty is added to encode the desired registration. The approach can be used in a wide variety of applications for example, the authors show how to detect the changes in the heart shape at different time points of the cardiac cycle.

Another segmentation method is presented by Genctav and Tari in Chapter 8. The authors investigate an approach for multi-parameter Mumford-Shah segmentation. Mumford-Shah image segmentation is a widely used approach that combines data fidelity with regularization term to classify the image pixels into foreground and background. The algorithm requires parameter tuning and multiple parameter sets yield different segmentation results. Genctav and Tari combine the multiple solutions into a feature vector that is obviously richer than using a single representation. Then they use dimensionality reduction to project the feature vector into a lower dimensional space with the highest information content. Such solution provides a much better segmentation due to information integration.

Image deblurring is another application presented in Chapter 9 by Bardsley and Howard. Image deblurring refers to the process of making images sharp and focused. It can be very useful in restoring old images. Image deblurring is usually modeled as an ill-posed linear inverse problem. The authors show to obtain a well-posed formulation by adding an L1-penalty term to the negative-log likelihood. Then, they formulate the minimization of the energy function as a bound and constrained optimization problem and solve it using iterative augmented Lagrangian.

Last but not least, Chapter 10 discusses shape patterns in digital fabrication, a comprehensive survey conducted by Yılmaz et al. on negative Poisson's ratio metamaterial.

Put together, the chapters of these proceedings shed the light of a wide variety of shape analysis methods and their applications to different image processing problems.

Princeton, NJ, USA   Noha El-Zehiry

Contents

1 Medial Fragments for Segmentation of Articulating Objects in Images 1
Erin Chambers, Ellen Gasparovic, and Kathryn Leonard

2 Scaffolding a Skeleton 17
Athina Panotopoulou, Elissa Ross, Kathrin Welker, Evelyne Hubert, and Géraldine Morin

3 Convolution Surfaces with Varying Radius: Formulae for Skeletons Made of Arcs of Circles and Line Segments 37
Alvaro Javier Fuentes Suárez and Evelyne Hubert

4 Exploring 2D Shape Complexity 61
Erin Chambers, Tegan Emerson, Cindy Grimm, and Kathryn Leonard

5 Phase Field Topology Constraints 85
Rüyam Acar and Necati Sağırlı

6 Adaptive Deflation Stopped by Barrier Structure for Equating Shape Topologies Under Topological Noise 95
Asli Genctav and Sibel Tari

7 Joint Segmentation and Nonlinear Registration Using Fast Fourier Transform and Total Variation 111
Thomas Atta-Fosu and Weihong Guo

8 Multi-parameter Mumford-Shah Segmentation 133
Murat Genctav and Sibel Tari

9 L^1-Regularized Inverse Problems for Image Deblurring via Bound- and Equality-Constrained Optimization 143
Johnathan M. Bardsley and Marylesa Howard

10 Shape Patterns in Digital Fabrication: A Survey on Negative Poisson's Ratio Metamaterials 161
Bengisu Yılmaz, Venera Adanova, Rüyam Acar, and Sibel Tari

Chapter 1
Medial Fragments for Segmentation of Articulating Objects in Images

Erin Chambers, Ellen Gasparovic, and Kathryn Leonard

Abstract The Blum medial axis is known to provide a useful representation of pre-segmented shapes. Very little work to date, however, has examined its usefulness for extracting objects from natural images. We propose a method for combining fragments of the medial axis, generated from the Voronoi diagram of an edge map of a natural image, into a coherent whole. Using techniques from persistent homology and graph theory, we combine image cues with geometric cues from the medial fragments to aggregate parts of the same object into a larger whole. We demonstrate our method on images containing articulating objects, with an eye to future work applying articulation-invariant measures on the medial axis for shape matching between images.

1.1 Introduction

This paper is a first step toward bringing the power of the Blum medial axis [2], a power amply demonstrated on pre-segmented shapes [13, 14, 18], to apply to shapes in natural images. The medial axis has excellent properties for representing 2D shape. It captures both region and boundary information, encodes shape parts and their relationships in its branching structure, and represents local feature size in the radius function. Unfortunately, its properties are only well-defined for closed shape contours, which often are difficult to extract from natural images in an automated way.

E. Chambers (✉)
Department of Computer Science, St Louis University, St Louis, MO, USA
e-mail: erin.chambers@slu.edu

E. Gasparovic
Department of Mathematics, Union College, Schenectady, NY, USA
e-mail: gasparoe@union.edu

K. Leonard
Occidental College, Department of Computer Science, Los Angeles, CA, USA
e-mail: kleonard.ci@gmail.com

© The Author(s) and the Association for Women in Mathematics 2018
A. Genctav et al. (eds.), *Research in Shape Analysis*, Association for Women in Mathematics Series 12, https://doi.org/10.1007/978-3-319-77066-6_1

1

This paper proposes a method for using the output of a standard segmentation algorithm, n-cuts [16], to aggregate segmented regions in natural images into larger objects whose medial axes capture important shape features. As a target application, we focus on images of shapes with articulating parts. Recent progress on articulation-independent shape understanding [8] depends upon measurements taken along the Blum medial axis of a shape. Those results motivate the need for extracting a stable medial axis that captures sufficient part information, so that the techniques of [8] may be applied. While many segmentation algorithms perform reasonably well, they often segment parts of the same object into different regions. We show that the structure of the medial axis can help recover more of the shape whole.

Object detection based on medial/skeletal shape models from pre-segmented shapes has been well-researched [18]. In [1], the authors employ a novel shape recognition method that instead relies on extracting the medial axis from an unsegmented arbitrary image. This paper expands that work in a new direction, computing the medial axes of regions obtained from a segmentation algorithm that is known to oversegment. We connect subsets of these medial fragments based on image and proximity information and then refine the resulting medial structure to obtain a valid medial axis for a shape containing multiple segmented regions. In Section 1.2 we discuss the medial axis and associated measures. Section 1.3 describes our method. Section 1.4 discusses our results, including some forward-looking examples that show the promise of using measures of shape tubularity (ST) [15] and erosion thickness (ET) [10] to provide successful part similarity measures for articulating shapes.

1.2   Background and Related Work

The *Blum medial axis* of a simple, closed plane curve, or shape, is the closure of the locus of maximally inscribed circles, together with their radii [2]. It is well-known [3, 12] that for sampled curves in 2D, the Voronoi vertices of the sampled points and their respective distances converge to the medial axis points and radii as the sampling becomes infinitely dense. The medial axis captures the intuitive structure of a shape, but often has been viewed as impractical for images because of its sensitivity and the perceived need for a pre-segmented shape.

The *extended distance function*, or EDF, was first introduced as a way to prune the medial axis and for applications such as shape alignment and description [10]. Intuitively, EDF measures the radius of the longest "tube" which fits in the interior of a 2D shape and extends Blum's original grassfire analogy for computing the medial axis [2]. In [8], the authors describe an automated parts decomposition for pre-segmented shapes based on the changes in values of a weighted version of the EDF, or WEDF, at medial axis branch junctures. The output of the algorithm is the number of levels of parts in a given shape, from coarse to fine, as well as which subtrees of the medial axis belong to each level. Because the parts' subtrees themselves can be quite complicated, the authors introduce the "trunk" of a subtree, which is a single curve traversing the subtree defined by the continuity of EDF values across branch points. These trunks represent the main supporting structures of the subtrees. Furthermore, the trunks are naturally tuned to the innate scales within the shape: a new trunk appears at its subtended part's coarsest scale, and a trunk at a coarser scale will contain trunks from finer scales.

Related to EDF and WEDF, *erosion thickness* (ET) at a point x is the difference between the EDF and local feature size for x. In a sense, ET measures the amount of the shape that is lost if the medial axis branch is pruned away. Similarly, *shape tubularity* (ST) is defined to be ET divided by EDF. Note that ST varies between 0 and 1, with 0 occurring only on the boundary of the medial axis. Intuitively, these measures are excellent options for dealing with articulation in an image since two trunks with the same ET and ST values will always describe the same part up to articulation [8].

1.3 Methods

Our preprocessing draws from standard smoothing and segmentation algorithms. Given an input image f, we begin by obtaining a smoothed version u of f as in [11], which requires minimizing the energy

$$E(u) = \int_{\Omega} (f - u) \, dx + \lambda \int_{\Omega} |\nabla u| \, dx. \tag{1.1}$$

Here, Ω denotes the image domain and $\lambda \in \mathbb{R}_{>0}$ is a weighting factor. We feed the smoothed image u into the n-cuts segmentation algorithm [16]. The resulting segmentation is an approximate solution to a graph partition problem, where an image is represented as a graph with pixels as nodes and edges weighted via an affinity matrix (Figure 1.1).

Fig. 1.1 Result of the n-cuts segmentation algorithm, in green, on an image of an elephant

We then extract preliminary medial information as in [1]. The output of the segmentation algorithm is a set of edges in the image, from which we compute the associated Voronoi vertices and edges. Whereas [1] then attempts to match information from a template medial axis into the messy Voronoi graph, we instead extract salient medial fragments from the Voronoi graph which we stitch together into a tree, producing a proposed medial axis of the shape.

There are many ways to determine affinity for joining nearby medial fragments. Here, we choose color, which provides a basis for decomposing the Voronoi points into clusters that help determine which fragments should be joined. The clustering methodology we use is based on the theory of *persistent homology*, which is a growing tool in computational topology used to identify prominent topological features and multi-scale connectivity information; see, e.g., [5]. There is a body of work that uses persistent homology to cluster, often combined with other information such as color to improve the segmentation [7, 9, 17].

Specifically, our clustering method involves persistent homology in dimension zero, which in our context provides a way of making sense out of the notion of the number of connected components of a point cloud. Given a set X of input points and a value $\epsilon \geq 0$, consider the family of balls of radius ϵ, $B_\epsilon(X) = \{B_\epsilon(x)\}_{x \in X}$. One may construct a simplicial complex on this set, known as the *Čech complex*, by adding a k simplex $x_0, x_1, \ldots, x_k$ if and only if $B_\epsilon(x_0) \cap \cdots \cap B_\epsilon(x_k) \neq \emptyset$. By incrementally increasing the value of ϵ, the point cloud thickens gradually and distinct components start to become connected in the associated complex. This process provides a filtration of the ambient space and enables one to keep track of the number of components one has at each scale ϵ. Each component is *born* at $\epsilon = 0$, and all but one will *die* at some later value of the scale parameter ϵ when it merges with another component in the complex. Eventually, for a large enough scale parameter, one will end up with a single connected component, and this component that never dies is said to have *infinite lifetime*.

For each point in the initial point cloud, we plot a point on the y-axis with y-coordinate equal to the death time of the corresponding component (where the death time is allowed to be infinity). The multi-set of these points is known as a *zero-dimensional persistence diagram*. The fundamental idea of zero-dimensional persistence is that points that are far from the diagonal are more likely to represent true features of the underlying space, while points close to the diagonal are more likely to be noise. One may also look at the gaps between successive points in the diagram to make inferences as to the number of connected components in the underlying point cloud. For example, for the point cloud in Figure 1.2 (a), most clustering algorithms would determine there to be three clusters. For certain scale choices, only two clusters may be reported (when the bottom two clusters are grouped together), and of course, for a sufficiently large scale, the entire point cloud may be treated as a single component. These options are captured in the persistence diagram in (b), although the point representing the single connected component with infinite lifetime (i.e., after all of the components have merged) is not pictured in this diagram. Indeed, two of the points on the y-axis are separated from the clump of noisy points near the origin. The y-values of those two points (their "death times") are equal to half the distance between the two bottom clusters and half the distance

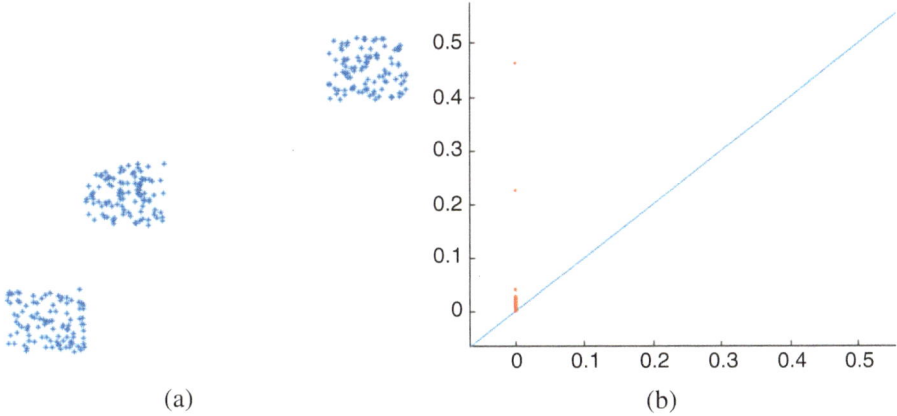

(a)   (b)

Fig. 1.2 Example illustrating zero-dimensional persistent homology

between the rightmost cluster and the middle cluster. Since there are three points in the persistence diagram (including the infinite lifetime point) that are separated from the rest of the pack near the origin, one may interpret the diagram to conclude that the point cloud contains three "persistent" connected components.

We now describe how we use zero-dimensional persistence in our clustering procedure. Our goal is to automatically determine how many LAB color clusters there are in a given image. To that end, we extract pixel LAB color values at the n unique Voronoi vertices. We then use this information to construct an $n \times 3$ point cloud in LAB color space. We perform zero-dimensional persistent homology on the point cloud and record the results in a persistence diagram. (Note that it is sometimes necessary to randomly downsample the number of points in the point cloud in order to be able to run persistence.) Then, we locate the largest gap between successive points in the persistence diagram, the idea being that the number of points above that gap (including the point in the diagram corresponding to the component with infinite lifetime) is an indication of the number of clusters in the point cloud in LAB space. Therefore, the number of clusters is taken to be the number of points above the largest gap between successive points in the persistence diagram. There is one caveat: we also require that the number of clusters be greater than two for better discriminative purposes. If the largest gap results in a cluster number of two, we move on to the second largest gap instead and take the number of points above that second gap to be the number of clusters.

For example, for the LAB image of the polar bear in Figure 1.3, the two largest gaps in the persistence diagram in Figure 1.4 occur between the first and second points down from the top and between the third and fourth points. Factoring in the dot corresponding to the component that never dies (plotted at the point $(0, -1)$ in the diagram), this means that the persistence-determined number of clusters should be either two or four. Since we require that there be more than two clusters in a given image, the number of clusters for the Voronoi vertices of this image is then determined to be four.

Fig. 1.3 Image of a polar bear displayed in LAB color

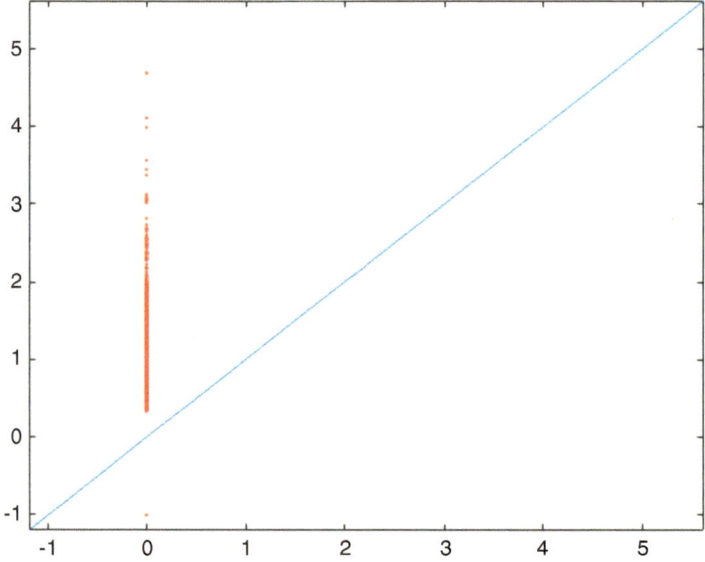

Fig. 1.4 Persistence diagram associated to the polar bear image, capturing the persistent homology of the point cloud of Voronoi vertices in LAB color space. Note that the number of clusters should be either two or four

The next step is to compute the induced subgraphs on each cluster, which may have multiple connected components. We join components that are spatially proximate by inserting an edge between the two closest Voronoi points in the two components when those points are at least as close as the length of the longest edge in either component. This often results in an invalid medial axis for a simply connected shape, where boundary points are found in the interior of the shape. To remove these points, and any associated loops in the medial axis, we trace the outer face of the medial graph and detect cycles or loops of length greater than two. For each loop, we identify the boundary points generating the loop, delete those on the interior of the loop, recalculate medial axis points from the remaining boundary points, and connect the resulting subgraph (now a tree) to the original medial axis.

For many applications, if one does not need to utilize the geometry of the boundary curve, the resulting medial axis is sufficient. For others, however, it may be necessary to relate the geometry of the medial axis with that of its boundary. In this case, we may want a set of ordered boundary points that corresponds to the extracted medial axis, which requires information regarding the connections or adjacencies between the associated points on the boundary. We order the boundary points by combining information about the proximity of other boundary points and the geometry of the medial axis. Boundary edges between points that are close but cross an edge of the medial axis are discarded in favor of edges between more distant points that do not cross. We also ensure that adjacent boundary points correspond to medial points that are geodesically close within the medial axis. Computing the medial axis of the resulting boundary may result in an improved segmentation of the shape.

To illustrate our method and motivate the discussion on our results, we now focus on a particular example of an image of a dancer. We demonstrate the outcome of each step of our method for this example, displaying the results in Figure 1.5. The result of the n-cuts segmentation of this image is given in (b) of that figure. Next, the dancer is shown in LAB color in (c), with the Voronoi points computed from the n-cuts segmentation edges displayed in color according to cluster number in (d). Part (e) of the figure shows the Voronoi graph of the body of the dancer, corresponding to the blue cluster in (d), obtained by computing the outer face of the Voronoi graph for that cluster. The cycles are highlighted in magenta, and part (f) provides a zoomed-in view of three of the loops. The next step is the recomputation of medial axis subgraphs to remove the loops, the results of which may be seen in parts (g) and (h). Finally, after finding the correct ordering of boundary points that correspond to the extracted medial axis, we compute the medial axis of the resulting boundary. The end product is the segmentation of the dancer in Figure 1.5 (i).

1.4 Results and Discussion

Our preliminary results using a small database of articulating shapes in images are promising. For example, the medial fragment-based segmentations obtained as a result of the process described in the previous section for an image of a black panther on a white background and a polar bear on a snowy background are pictured in Figure 1.6. In the case of the dancer in Figure 1.7, the leftmost segmentation was obtained using our ordinary procedure outlined in the previous section. Notice that there are some missing triangles in the Delaunay triangulation as a result of the loop

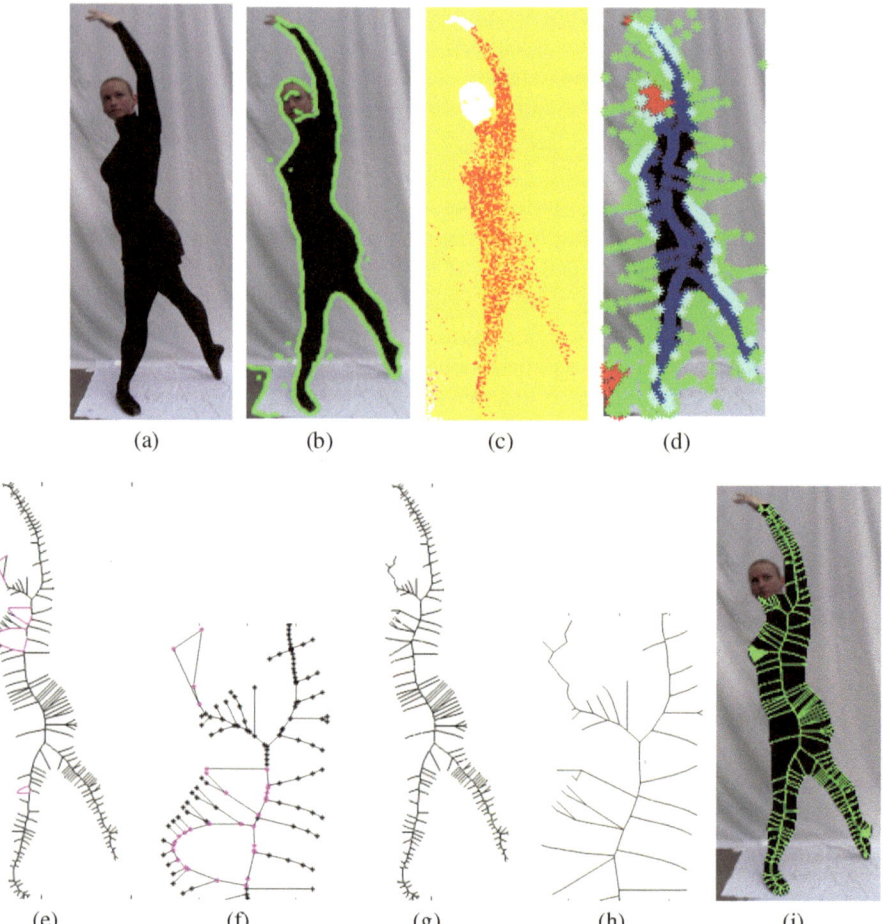

Fig. 1.5 (a) An image of a dancer and (b) the results of the $n-$cuts segmentation. (c) The LAB color image of the dancer. After computing zero-dimensional persistent homology on the Voronoi vertices in LAB color space, the automatically determined cluster number was four. (d) Results of performing k-means clustering on the Voronoi vertices with $k = 4$. (e) The Voronoi graph of the body of the dancer (corresponding to the blue cluster in (d)) with cycles highlighted in magenta. (f) A closer view of three of the loops in the graph. (g) The medial axis (and (h) a zoomed-in view) of the dancer after our procedure to remove loops and recompute a medial axis structure. (i) Final dancer segmentation, including the improved segmentation procedure at the end of Section 1.3

medial axis replacement procedure. The improved segmentation in the middle (and overlaid on the original image on the right) was obtained as a result of determining the correct set of ordered boundary points associated to the medial fragment points, as mentioned at the end of Section 1.3, and then recomputing the Blum medial axis of the resulting boundary. Recall that the segmentation of the dancer in Figure 1.5 was constructed using this improved segmentation process, as well.

(a)

(b)

(c)

Fig. 1.6 Three additional examples of our medial fragment-based segmentation process

In Figure 1.8, we see an image of a horse and the resulting segmentation that we obtained. Note that it is an oversegmentation of the horse, in part owing to the fact that we picked up Voronoi points belonging to the trees in the background, since the Voronoi points of the trees in this example were assigned to the same cluster as the one containing the medial points within the body of the horse.

Our methods exhibit potential utility for combining parts of the same object into a coherent whole that were put into distinct regions by the initial segmentation. This idea is illustrated in Figures 1.9 and 1.10. First, in Figure 1.9, the initial n−cuts segmentation appears to separate the woman's leg in the image into a separate region from the rest of her body. However, our clustering procedure aggregated the Voronoi

Fig. 1.7 (a) Initial medial fragment-based segmentation of a dancer in an image. (b) Improved segmentation of the dancer, which requires the determination of the set of ordered boundary points associated to the medial fragment points. (c) The improved segmentation overlaid on the original image

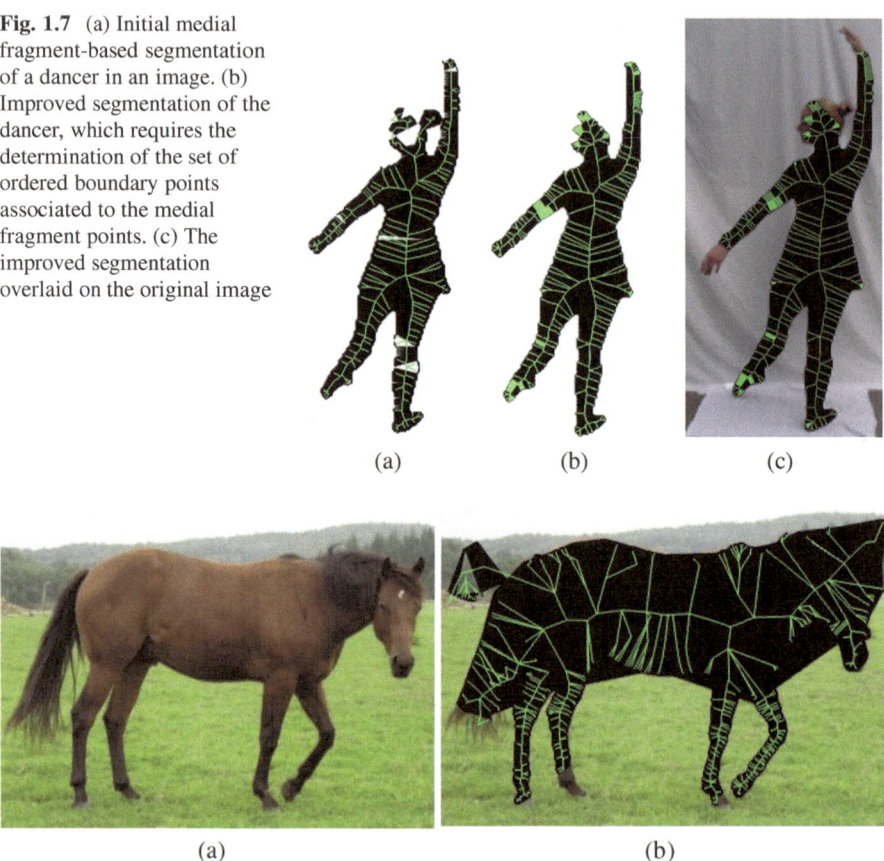

(a)   (b)   (c)

(a)   (b)

Fig. 1.8 Our segmentation (b) of the horse in the image in (a)

points located inside the woman's body into the same cluster, and therefore the triangulation we obtain from the largest connected component of the subgraph on that cluster — namely, the medial fragment shown in green in Figure 1.9 (b) — yields the given segmentation. We point out that, if the medial fragments capturing the ankles and feet of the woman in the image were not separate components in the cluster subgraph but were instead joined to the primary medial fragment for the shape, the overall segmentation would be improved.

Second, in Figure 1.10, we see an image of a whaletail in the ocean with non-uniform coloring on the tail. As a result of this color variation, the $n-$cuts segmentation output contains a large amount of extraneous detail inside the main boundary of the whaletail. Our procedures of clustering by color, joining spatially proximate components of the induced Voronoi subgraph within a particular cluster, removing loops, and reforming connections yield the coherent final segmentation in Figure 1.10 (c).

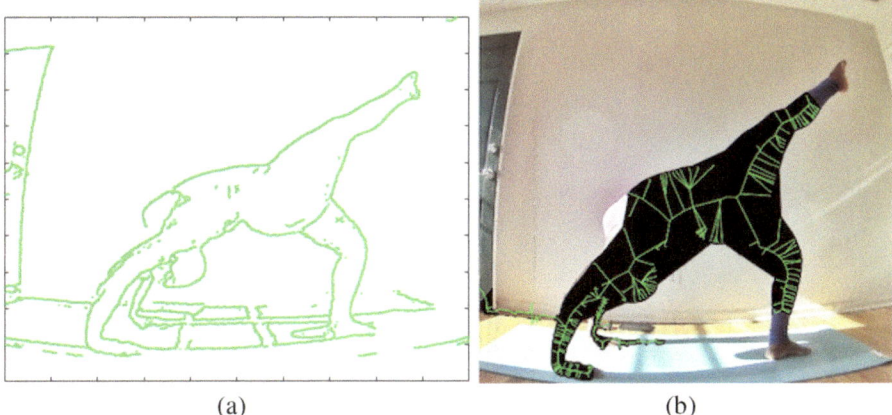

(a)   (b)

Fig. 1.9 (a) The n−cuts segmentation of a woman in a yoga pose. Notice that the edges in the segmentation put parts of the same object into different regions (viz., one of the legs of the woman and the rest of her body). (b) Our medial fragment-based segmentation of the woman in the image. Note that our procedure combines the separate parts in the segmentation into one coherent object

Indeed, in future work, we hope to increase the robustness of our method. Our persistent homology clustering is based solely on image color at the Voronoi points. This naturally works well on objects with fairly uniform color, but is less successful when the input image has variations in color. Other affinity measures, such as texture or statistical moments, considered within a Voronoi cell are likely to improve our ability to join fragments from regions within the same shape with large color variation. In addition, replacing a loop somewhere in the medial axis and then rejoining to the rest of the medial axis can damage the structure of the medial axis overall once it is joined to the larger medial structure. Inserting a check for violation and determining an alternative, non-violating subgraph would improve our final skeleton. Finally, drawing on work to understand medial structures in regions with multiple objects [4, 6], we hope to analyze the interaction between disjoint medial structures to better determine when two regions are likely to belong to the same object.

In ongoing work, we plan to extract the ET and ST values along trunks from each of the medial fragments in an image, from all parts at all levels. Each image is then represented by an unknown number of vectors of ET and ST values. We plan to employ the bag-of-words approach, using k-means clustering on the ET/ST vectors from the entire training set, to obtain feature vectors for each image that can be used for parts matching between images. As shown in Figures 1.11 and 1.12, the ET and ST values for similar parts in different images are similar; this motivates the use of these measures as a powerful tool for matching articulated shapes, where even partial or obstructed images will have a high likelihood of matching values along the parts.

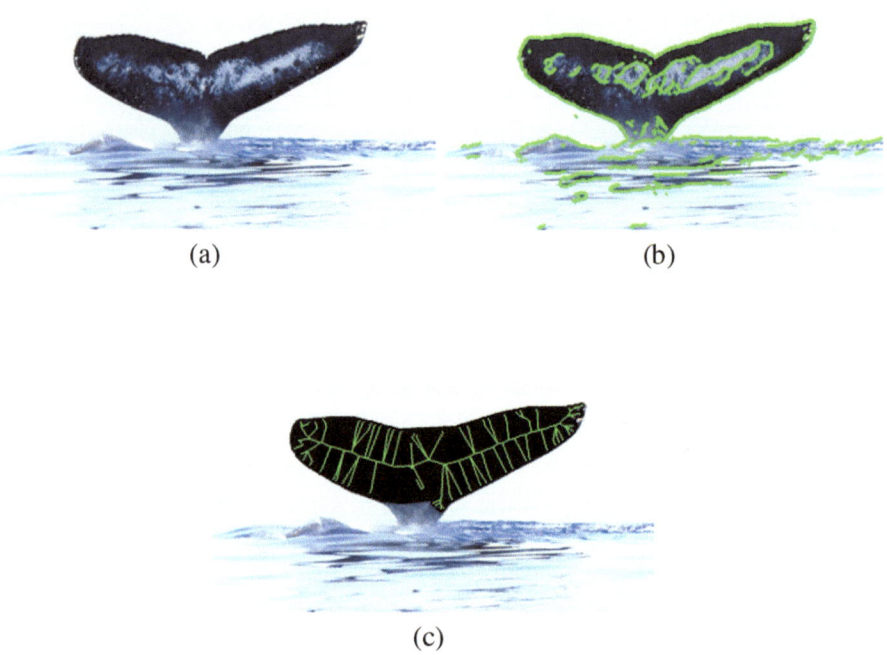

Fig. 1.10 (a) An image of a whaletail and (b) the result of the *n*-cuts segmentation. Notice that the nonuniform coloring on the whaletail led to a large amount of detail showing up in the segmentation inside the main boundary of the whaletail. (c) Our medial fragment-based segmentation finds the coherent object in the image

We end with a word of caution. Our original goal in this work was to extract a nice, ordered boundary curve for each of the segmented shapes. After extensive trial and error using boundary point proximity and geometry together with medial point adjacency and geometry to infer which boundary points should be joined by an edge, we have concluded that no one method can successfully extract such a boundary. Because of the noisiness of the edges extracted from the image, and the resulting noisiness of the medial fragments, no one boundary adjacency inference method was able to satisfy the requirements for boundaries of all objects; rather, each method succeeded for some images but broke on others. The dearth of publishing

Fig. 1.11 *Top row*: Medial axis and erosion thickness (ET) values displayed as a heat map for different dancer images. *Bottom row*: Close-up of ET values in heat maps for the legs. Note the similar coloring between images

about extracting valid medial axes from unsegmented objects in natural images may suggest that many researchers have spent time unsuccessfully tackling this project. It may be that probabilistic methods are required, an area we did not explore, but it may also be that striving for the best possible segmentation before extracting medial information is the most desirable approach.

Fig. 1.12 *Top row*: Medial axis and Shape Tubularity (ST) values displayed as a heat map for different dancer images. *Bottom row*: Close up of ST values heat maps for legs. Note the similar coloring between images

References

1. Bal, G., Diebold, J., Chambers, E.W., Gasparovic, E., Hu, R., Leonard, K., Shaker, M., Wenk, C.: Skeleton-based recognition of shapes in images via longest path matching. In: Research in Shape Modeling, pp. 81–99. Springer, Cham (2015)
2. Blum, H.: A transformation for extracting new descriptors of shape. In: Models for the Perception of Speech and Visual Form, pp. 362–80. MIT Press, Cambridge, MA (1967)
3. Brandt, J.: Convergence and continuity criteria for discrete approximations of the continuous planar skeleton. CVGIP: Image Underst. **59**(1), 116–124 (1994)
4. Damon, J., Gasparovic, E.: Medial/skeletal linking structures for multi-region configurations. Memoirs Am. Math. Soc. (2014). http://arxiv.org/abs/1402.5517v2
5. Edelsbrunner, H., Harer, J.: Computational Topology: An Introduction. American Mathematical Society, Providence, RI (2010)
6. Gasparovic, E.: The Blum medial linking structure for multi-region analysis. Ph.D. Thesis (2012)

7. Ge, Q., Lobaton, E.: Consensus-based image segmentation via topological persistence. In: The IEEE Conference on Computer Vision and Pattern Recognition (CVPR) Workshops, pp. 95–102 (2016)
8. Leonard, K., Morin, G., Hahmann, S., Carlier, A.: A 2D shape structure for decomposition and part similarity. In: International Conference on Pattern Recognition (2016)
9. Letscher, D., Fritts, J.: Image segmentation using topological persistence.In: Kropatsch, W.G., Kampel M., Hanbury A. (eds.) Computer Analysis of Images and Patterns. CAIP 2007. Lecture Notes in Computer Science, vol. 4673, pp. 587–595. Springer, Berlin (2007)
10. Liu, L., Chambers, E.W., Letscher, D., Ju, T.: Extended grassfire transform on medial axes of 2D shapes. Comput. Aided Des. **43**(11), 1496–1505 (2011)
11. Rudin, L.I., Osher, S., Fatemi, E.: Nonlinear total variation based noise removal algorithms. Phys. D **60**, 259–268 (1992)
12. Schmitt, M.: Some examples of algorithm analysis in computational geometry by means of mathematical morphological techniques. In: Proceedings of the Workshop on Geometry and Robotics, pp. 225–246. Springer, London (1989)
13. Sebastian, T.B., Kimia, B.B.: Curves vs. skeletons in object recognition. Signal Process. **85**(2), 247–263 (2005)
14. Sebastian, T.B., Klein, P.N., Kimia, B.B.: Recognition of shapes by editing their shock graphs. IEEE Trans. Pattern Anal. Mach. Intell. **26**(5), 550–571 (2004)
15. Shaked, D., Bruckstein, A.M.: Pruning medial axes. Comput. Vis. Image Underst. **69**(2), 156–169 (1998)
16. Shi, J., Malik, J.: Normalized cuts and image segmentation. IEEE Trans. Pattern Anal. Mach. Intell. **22**(8), 888–905 (2000)
17. Skraba, P., Ovsjanikov, M., Chazal, F., Guibas, L.: Persistence-based segmentation of deformable shapes. In: IEEE Computer Society Conference on Computer Vision and Pattern Recognition Workshops (CVPRW) (2010)
18. Trinh, N.H., Kimia, B.B.: Skeleton search: category-specific object recognition and segmentation using a skeletal shape model. Int. J. Comput. Vis. **94**(2), 215–240 (2011)

Chapter 2
Scaffolding a Skeleton

Athina Panotopoulou, Elissa Ross, Kathrin Welker, Evelyne Hubert, and Géraldine Morin

Abstract The goal of this paper is to construct a quadrilateral mesh around a one-dimensional skeleton that is as coarse as possible, the "scaffold." A skeleton allows one to quickly describe a shape, in particular a complex shape of high genus. The constructed scaffold is then a potential support for the surface representation: it provides a topology for the mesh, a domain for parametric representation (a quad-mesh is ideal for tensor product splines), or, together with the skeleton, a grid support on which to project an implicit surface that is naturally defined by the skeleton through convolution. We provide a constructive algorithm to derive a quad-mesh scaffold with topologically regular cross-sections (which are also quads) and no T-junctions. We show that this construction is optimal in the sense that no coarser quad-mesh with topologically regular cross-sections may be constructed. Finally, we apply an existing rotation minimization algorithm along the skeleton branches, which produces a mesh with a natural edge flow along the shape.

A. Panotopoulou (✉)
Dartmouth College, Hanover, NH, USA
e-mail: athina@cs.dartmouth.edu

E. Ross
MESH Consultants Inc., Fields Institute for Research in the Mathematical Sciences, Toronto, ON, Canada
e-mail: elissa.ross@meshconsultants.ca

K. Welker
Trier University, Trier, Germany
e-mail: welker@uni-trier.de

E. Hubert
INRIA Méditerranée, BP 93, Sophia Antipolis, France
e-mail: evelyne.hubert@inria.fr

G. Morin
Toulouse Institute of Computer Science R, Toulouse, Garonne (Haute), France
e-mail: morin@n7.fr

© The Author(s) and the Association for Women in Mathematics 2018
A. Genctav et al. (eds.), *Research in Shape Analysis*, Association for Women in Mathematics Series 12, https://doi.org/10.1007/978-3-319-77066-6_2

2.1   Introduction

A skeleton, like the skeleton of vertebrates, defines synthetically a shape and makes intelligible its general organization while helping to visualize its possible deformations. Mathematically, a skeleton is a union of lower-dimensional manifolds that is homotopy equivalent to the shape. As such it characterizes the topology of the shape. Skeletons, and more particularly medial representations, are computed for analyzing and identifying shapes, with a purpose of automatic recognition of objects, or of a human eye diagnosis. Such models lend themselves to identification and matching and are therefore used for retrieval [6, 17].

Computer graphics also use skeletons as a starting point for shape design [10]. Skeletal animation (rigging) is an established technique to animate moving characters. Ultimately, shapes will be represented by their boundary surfaces, as either an implicit or parametric surface, or as a fine mesh. Yet the overall design of the shape and its topology is more easily entered as a skeleton. Our approach thus allows to create a two-manifold mesh of surfaces of high genus starting from a schematic description of the shape (the skeleton) rather than plugging handles on an existing mesh as in [15], for example.

The aim of our project is to construct a quadrilateral mesh around a skeleton that is as coarse as possible, like a scaffold on which to build the surface. Such a mesh provides the first step for defining a joint surface representation for a skeleton-based shape; the scaffold quads can be refined by subdivision to fit a regular quad-mesh to a surface shape, as in [19]. Our construction is optimal in the sense that it only consists of quads, in contrast to other methods of constructing meshes from skeletons [2, 11, 19]. In addition, our method is minimal in the sense that it is not possible to construct a coarser quad-mesh of the same skeleton with uniform sections. By uniform sections, we mean that every section is of the same type of polygon, for example, triangle, quad, and hexagon.

The paper is organized as follows. In Section 2.2, we give an overview of the related work. In Section 2.3, we define the skeleton we shall work with and outline the construction of the scaffold. The two main parts of the algorithm are then described in the next two sections. Finally, in Section 2.6 we present a selection of examples of skeleton for which we generate a scaffold with the described method.

2.2   Related Work

3D Skeletons   The skeleton (medial axis and radius function on the axis) can be seen as a dual to a surface representation which captures all shape aspects that an explicit or implicit representation captures [18]. Here, we do not consider a skeleton that characterizes the entire shape, but rather a simple piecewise linear skeleton, similar to skeletons used for animation [14], or for convolution surfaces [4, 8, 9, 12].

Inverse Skeletonization, Garbing, and Skeletally Driven Modeling Recent research works have tackled the problem of generating an explicit mesh based on a given skeleton. Bærentzen et al. [2] presented an algorithm to create a quad-dominant polygonal mesh. This method, similar to our work, starts by triangulating the spheres on vertices with three or more neighbors (branch nodes). Our approach uses quadrilaterals as building blocks instead of triangles, which results in a more regular mesh. Their method is intended for simpler tree structured meshes, as they do not take into account torsion of the skeleton (i.e., how far the skeleton is from planar) after the rotation step. Furthermore, they also do not bound the number of vertices of a face and may have sections with a varying number of vertices. Usai et al. [19] use only quadrilaterals for the branching nodes and consider torsion, but their construction leads to meshes with irregular sections and to ambiguous final representations in an effort to solve the algorithm's inherent T-junctions.

Bærentzen et al. [3] proposed the use of a polar-annular mesh as a lightweight representation of any polygonal mesh that is accompanied by a skeletal representation. Because their technique may produce meshes with high valency polar triangle fans, it may be more memory intensive than ours and it lacks the explicit encoding of radius on the skeleton. They do not provide a theoretical justification of the completeness for the meshing of branch points. Therefore, our proposed construction could complement their definition of the skeleton-mesh co-representation, as our algorithm has a complexity linear in terms of the number of vertices in the graph. Ji et al. [11] mesh the branches before the branching nodes, to produce a quad-dominant mesh. Their algorithm can create triangles and irregular vertices on the joints. Also, in cases that they produce meshes of bad edge flow (how well the lines of the edges appear to be aligned with the shape of the skeleton), they need to solve a quadratic optimization problem. In other words, the method used in [11] for selecting the local frames may be heuristic at every node. In contrast, we use a heuristic approach for the frames at the branch nodes, but the frames on the remaining nodes are defined using rotation minimization. This allows for the straightforward realization of examples such as the spiral in Figure 2.6. The similar problem addressed in [16] is different in that it imposes at the outset a given polygonal section to the branches of the skeleton.

2.3 Construction Outline

In this section we first define a skeleton as an embedding in $\mathbb{R}^3$ of a simple graph and then split our construction of a scaffold into two major steps. These two steps are detailed in the next two sections.

A *skeleton* is defined by its topology and its geometry. The topology is given by a simple, unweighted, undirected, and connected graph $G = (V, E)$. Here V denotes the set of vertices and E denotes the set of edges. Each vertex v of V is associated with a point p_v in $\mathbb{R}^3$ and a radius $r_v > 0$ which defines a sphere centered at p_v (Figure 2.1). In other words, a *skeleton* is defined as the triple (G, p, r), where

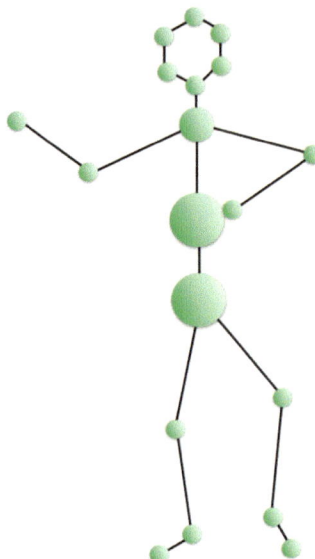

Fig. 2.1 A skeleton: an embedded graph with a sphere at each vertex

- $G = (V, E)$ denotes the graph defined above
- $p: V \to \mathbb{R}^3$,

  $v \mapsto p_v$ is the graph embedding, i.e., an assignment of unique positions from $\mathbb{R}^3$ to each vertex of the graph, and
- $r: V \to \mathbb{R}^+$,

  $v \mapsto r_v$ with r_v denoting the radius associated to the vertex $v \in V$.

Consequently, any edge of the graph G is embedded as a line segment of $\mathbb{R}^3$. Edges are assumed to intersect only at end points. We furthermore assume that, outside the spheres centered at the vertices, the line segments are far enough from each other for the scaffold not to intersect. In the present work, we also assume that the spheres defined on the embedded graph do not intersect. There does not seem to be a large obstacle to treating this case, as shown in some of the examples (see Figure 2.8). Yet understanding how the intersections of the spheres could cause failures in the resulting mesh is a topic of future work.

We distinguish between the so-called joints, branch points, and extremities: the image in $\mathbb{R}^3$ of a vertex of valency 2 (respectively 1) is called a *joint* (respectively an *extremity*). A *branch point* is the image p_v of a vertex v of valency 3 or more. Together, we call the joints extremities and branch points *skeleton nodes* (or simply *nodes*, where it is clear from context).

A *branch* is a subgraph connecting either two branch points or a branch point to an extremity, with no intermediate branch points.

For a given skeleton (V, E, p, r), a *scaffold* is a closed orientable polyhedral surface enclosing the union of line segments defined by the skeleton, while being homotopically equivalent to it. The scaffold we shall construct is a quad mesh, the

Fig. 2.2 The cross-section of
a branch sleeve. In this work
we describe a construction
method for quad meshes with
quadrilateral cross-sections
(*left*)

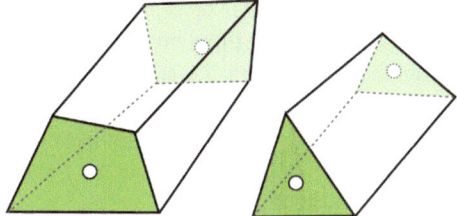

faces of which we call *scaffold quads* (white quads in Figure 2.2). Each scaffold
quad is associated with a single skeleton edge, and a skeleton edge is associated
with exactly four scaffold quads (a *sleeve*). Our construction of a scaffold for a
skeleton consists of two main steps:

Partition: For all branch points p_v, partition the sphere S_v centered at p_v and
with radius r_v, according to the issuing branches. In this step, the spheres are
partitioned into quadrilateral faces, called *spherical quads* (dark green quad in
Figure 2.2). Each spherical quad face is intersected by a single line segment
joining the branch point to a neighbor vertex.

Sleeve: With a depth-first search (DFS) order, starting from any branch point or
extremity, *sleeve* all the branches. In this step, we compute a propagated spherical
quad (light green quad in Figure 2.2) for every joint (*i.e.,* skeleton node of
degree 2) following the DFS order. We use a heuristic alignment for the joints or
extremities connected to a branch point and the rotation minimization algorithm
for all remaining skeleton nodes. We use these quadrilaterals surrounding the
nodes of the skeleton to mesh every edge of the skeleton using four quad faces.
Finally, we close the mesh at extremities.

2.4 Partitioning the Sphere at a Branching Point

In this section we detail the first step of the construction of a scaffold, the partition
of the sphere around a branch point of the skeleton into spherical polygons that
separate the adjacent edges of the skeleton. We look for a regular such partition, *i.e.,*
one where all the spherical polygons have the same number of vertices. We show
that the use of quadrangles is the simplest possibility that works independently of
the number of adjacent edges to the branch point. Note that we address here only
the partition of the sphere around the branch points as the joints and extremities are
dealt with when we treat the sleeves.

Let us consider one branch point, and we denote N its valency ($N \geq 3$ by
definition). We first compute N points on the sphere, corresponding to the N
adjacent edges of the skeleton. We want to partition the sphere into N regular
spherical polygons such that each spherical polygon contains exactly one of the
N points. Here, regular is meant in a topological sense, i.e., each spherical polygon

has an equal number of vertices. Each spherical polygon, also referred to as a face, will be the support of an outgoing sleeve (part of the scaffold) enclosing the skeleton branch starting with the adjacent edge stemming out this face. The vertices of the spherical polygons will be the only vertices of the scaffold that belong to this sphere. In order to *sleeve* a branch connecting two branch points, without introducing T-junctions, it is necessary that the faces around each branching point all have the same number of vertices.

We seek the simplest scaffold as possible. Hence, we first address the question of how the sphere can be partitioned such that the number of vertices of the faces is minimal. We first show that we cannot always obtain an appropriate partition of the sphere into spherical triangles. We then provide a construction to partition the sphere into spherical quadrangles. This proves that four is the minimal number of vertices that our spherical polygons should have to partition the sphere at branch points.

2.4.1 Minimum Number of Vertices of Faces on the Sphere

Let us consider a sphere centered at a branch point of the skeleton under consideration. Assuming that we have N points on the sphere, corresponding to the intersection of the sphere with the N outgoing branches of the skeleton, we wish to separate the sphere into regular faces, such that each region contains exactly one point. As we seek the simplest scaffold possible, our goal is to minimize the number of vertices of the faces, i.e., we should try to have triangles. However, we prove that partitioning the sphere into N triangles is not possible if N is odd. Essential to this proof is the Euler characteristic χ which is classically defined for polyhedral surfaces by

$$\chi = v - e + f, \tag{2.1}$$

where v, e, and f are the numbers of vertices, edges, and faces in the polyhedron under consideration. For more information about the Euler characteristic, we refer to the literature, e.g., [1, 7, 13]. In the case of a closed orientable surface, it can be calculated from its genus g by $\chi = 2(1 - g)$. The genus of a surface is equal to the number of holes of the surface under consideration. For the sphere g, it is equal to zero, i.e., we get

$$2 = v - e + f. \tag{2.2}$$

With the help of this formula, we can prove the following theorem.

Lemma 2.1 *The sphere S^2 cannot be partitioned into N triangles if N is odd.*

Proof Consider formula (2.2) for this setting. We have $f = N$ because we want to have N triangles. Moreover, we assume that each face has the same number k

of vertices; then $e = \frac{N \cdot k}{2}$, as each face has also k edges, but each edge is counted twice. Thus, we get

$$v = \frac{N \cdot k}{2} - N + 2 = N \left(\frac{k}{2} - 1 \right) + 2. \tag{2.3}$$

If all faces were triangles, then $k = 3$. From (2.3) we get $v = \frac{N}{2} + 2$. This is not possible if N is odd.   □

The lemma above states that we cannot partition the sphere into an odd number of triangles. However, it may be possible to partition any sphere into N quads, using $N + 2$ vertices. This is because, for quad faces, we have $k = 4$ in the above proof and we get $v = N + 2$ from (2.3). In the rest of this paper, we will construct such a partition and deduce a quad-mesh scaffold of a skeleton of arbitrary topology, with quadrilateral cross-sections (inherited from the quad faces on the sphere; see Figure 2.2). Because the mesh is based on the partition of spheres into quads, we avoid T-junctions by construction. Next, we give an actual constructive algorithm to partition the sphere.

2.4.2 Algorithm to Partition the Sphere into Quadrangles

Lemma 2.1 states that the sphere cannot be partitioned into N triangles if N is odd. However, we may construct a partition of the sphere into quads. This subsection is devoted to the formulation and explanation of an algorithm for such a sphere partition into quads (see also the pseudocode recorded in Algorithm 2.1).

The *input* of the algorithm is the center of a sphere, i.e., a branch point, and its radius, together with a list of the neighbor skeleton vertices of the branch point. The *output* of the algorithm is a set of quadruples of scaffold vertices, each quadruple defining a face in the partition of the sphere.

We now outline how the sphere can be partitioned inductively into quads. For this purpose, let (G, p, r) be a skeleton, and consider a branch point C of (G, p, r). Suppose first that C has three neighbor vertices $N_1, N_2, and N_3$. For each neighbor N_i of C, we build the half line starting at C in direction N_i. The intersection of each half line with the sphere gives one point P_i on the sphere (Figure 2.3a).

These three points uniquely determine two antipodal points Q and Q' on the sphere which are equidistant to P_1, P_2, and P_3. These may be computed as the intersection of the three medial planes between P_j and P_k, where $j, k \in \{1, \dots, 3\}$ with $j \neq k$. We connect Q and Q' with three geodesic curves. The geodesic between P_j and P_k is the half circle that results from intersection of the medial plane between P_j and P_k and the sphere, when P_j and P_k are two consecutive points on the sphere according to an angle projected on a plane orthogonal to $[QQ']$. The sphere is now partitioned like an orange with three arcs of great circles (cf. Figure 2.3b). Finally, we add one additional vertex on each geodesic at the midpoint of the geodesic passing between these two points. This creates a partition of the sphere into three (combinatorial) quadrilaterals (Figure 2.3c).

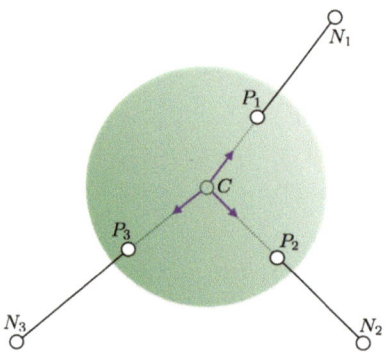

(a) A sphere representing an example of a branch point with three branches coming out of it

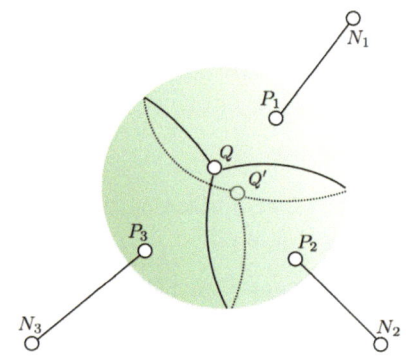

(b) Segmenting the sphere into spherical quads, each branch is included in a different quad

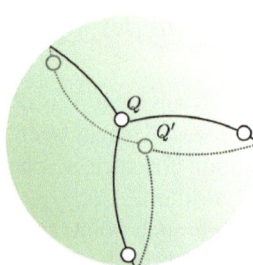

(c) The spherical quads on the surface of the sphere

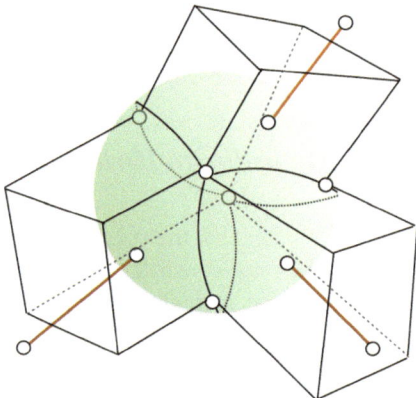

(d) The resulting scaffold mesh

Fig. 2.3 (**a**) A sphere with center C, neighbors $\{N_1, N_2, N_3\}$, and resulting points of intersection on the sphere $\{P_1, P_2, P_3\}$. (**b**) The segmentation of the sphere into sections. (**c**) The addition of points along the geodesics to create the quadrilaterals. (**d**) The scaffold mesh

Next assume that there are more than three points P_1, P_2, $and\,P_3$ on the sphere, i.e., we have additional edges starting at the vertex C. In this case, for each additional neighbor vertex N, let P be the intersection of the edge (C, N) with the sphere. P belongs to a quad face of the sphere, say the one of P_j. We place a vertex at the midpoint M of the geodesic between P and P_j and connect M to one of the diagonals of the quad. We choose the diagonal that offers the direction most orthogonal to $[P, P_j]$.

A relevant question to examine is how to best adjust the position of the quad vertices on the sphere. In our implementation, after introducing a new point P, we used the following heuristic to reposition each vertex q of a new quad containing

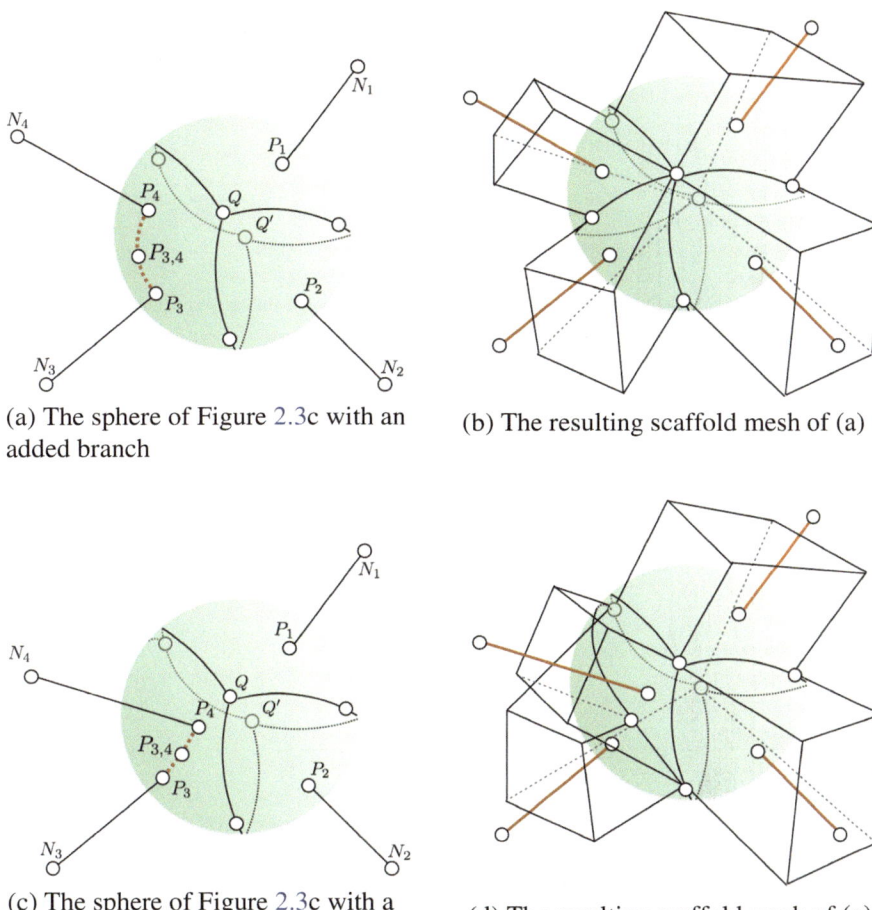

(a) The sphere of Figure 2.3c with an added branch

(b) The resulting scaffold mesh of (a)

(c) The sphere of Figure 2.3c with a different added branch

(d) The resulting scaffold mesh of (c)

Fig. 2.4 We represent the branch point as the green sphere; we use black-colored lines to represent spherical and scaffold quads and orange-colored lines to represent branches. (**a, c**) The fourth branch is in the same spherical face as branch 3. We take the midpoint of the geodesic connecting P_3 and P_4 to be the new vertex of the spherical quad faces surrounding P_3 and P_4. (**b, d**) The resulting meshes with an additional branch and corresponding new quad face on the sphere

P : we look at each quad on the sphere containing q. Each quad corresponds to a point P_i on the sphere, so the new position of q should be equidistant to all sphere points P_i. If q has just two neighbor quads, that is, there are two such P_is, we take the midpoint on the geodesic between P_{i_1} and P_{i_2}. If q belongs to three quads, that is, there are three P_is, we take the closest equidistant point (as shown above for Q in Figure 2.3); if there are more than three P_is, we take q equidistant to the P_is in a least square sense (cf. Figure 2.4). In the examples we treated, the so adjusted spherical quads remained an appropriate partition of the sphere: each face would

---

Algorithm 2.1 Partitioning branch point spheres

Data: (G, p, r) is a skeleton, $C \in \mathbb{R}^3$ is a branch point, r_C is the radius of the sphere at C,
$N = \{N_0, \ldots, N_m\}$ is the set of all vertex neighbors of C.

1: **function** SPHERE PARTITION(G, C)
2:     $S \leftarrow$ sphere of radius r_C centred at C
3:     $N_0, N_1, N_2 \in \mathbb{R}^3 \leftarrow$ the coordinates of the first three vertex neighbours of C in G
4:     **for** $i \leftarrow$ 0to2 **do**
5:         $P_i \in \mathbb{R}^3 \leftarrow$ intersection of half line from C to N_i with S
6:     **end for**
7:     $Q, Q' \in \mathbb{R}^3 \leftarrow$ the uniquely defined antipodal points on S that are equidistant to P_1, P_2, P_3
8:     **for** $i \leftarrow$ 0to2 **do**
9:         $g_i = (Q, Q') \leftarrow$ the geodesic separating P_i and $P_{i+1(\text{mod } 3)}$
10:        $P_{i,i+1} \in \mathbb{R}^3 \leftarrow$ intersection of g_i with the geodesic connecting P_i and $P_{i+1(\text{mod } 3)}$
11:    **end for**
12:    **for** $i \leftarrow$ 0to2 **do**
13:        $F_i = (Q, P_{i,i-1(\text{mod3})}, Q', P_{i,i+1(\text{mod3})}) \leftarrow$ the face of that contains P_i
14:    **end for**
15:    **for** each additional neighbor $N_i \in N, i \geq 3$ **do**
16:        $P_i \in \mathbb{R}^3 \leftarrow$ intersection of half line from C to N_i with S
17:        $F_j, j < i \leftarrow$ face of the quad-partitioned sphere containing P_i (note F_j contains P_j)
18:        $P_{i,j} \in \mathbb{R}^3 \leftarrow$ mid-point of the geodesic connecting P_i and P_j
19:        Connect $P_{i,j}$ with two non-adjacent vertices in the face F_j, and modify the face F_j
           accordingly (see discussion below).
20:    **end for**
21:    **return** partition of sphere into quads
22: **end function**

---

contain only one adjacent skeleton edge. It is nonetheless not clear how to prove such a method works in general. Other possible criteria of adjustment could regard minimizing the twist along the branches, or the convexity of the faces.

2.5   Constructing *Sleeves* Around Skeleton Edges

Algorithm 2.1 described a partition of each sphere into spherical quads. We now describe the "sleeving" procedure that will connect these partitioned spheres together and the partitioned spheres to the extremities (see Algorithm 2.2). Recall from Section 2.3 that a skeleton *branch s* is defined by a sequence of connected vertices $\{v_0, \ldots, v_l\} \in V$ such that v_0 is a *branch point*, v_l is either a *branch point* or an *extremity*, and v_k, for $0 < k < l$, is a joint. The sequences of nodes that describe the branches are recovered by a depth-first traversal starting from any extremity (see Section 2.3).

2.5.1 Algorithm: Creating a Sleeve Along the Branch

The idea behind creating a sleeve around a branch is to avoid creating a twist in the quadrilateral section, as we go along the branch. For this purpose, rotation-minimizing frames have been proposed (see, e.g., [5] for a review of recent minimizing frame methods) and require the definition of a frame attached to the branch.

2.5.1.1 Defining a Frame at v_0

The branch s corresponds in a unique way to a spherical quad (p^1, p^2, p^3, p^4) in the partition of the sphere around the branching point v_0. The first step is to associate to (p^1, p^2, p^3, p^4) a local orthonormal frame that approximates this quad in two ways. The frame should follow the direction of the branch and needs to approximate the plane defined by the corresponding quad, the unit vector that goes out of the branch point and is parallel to the first edge of the branch. In order to orient the axes x and y on the plane orthogonal to z, we use the following heuristic: we define the auxiliary vector $t = p^1 - p^2$ and compute the x-axis as $x_0 = z_0 \times t$ and the y-axis $y_0 = x_0 \times z_0$ (where $\times$ is the cross product). We define the aligned spherical quad $(q_0^1 = p^3, q_0^2 = p^4, q_0^3 = p^1, q_0^4 = p^2)$ and the orthonormal local frame (x_0, y_0, z_0) associated with the first vertex in the sleeve.

This original frame is then propagated along the branch. There are two possibilities: the branch either terminates with an *extremity* or with a *branch point*. This approach described above works well for all of our examples: the created frame is well oriented for the starting spherical quad, that is, no twisting happens on the set of starting faces (see Figure 2.5).

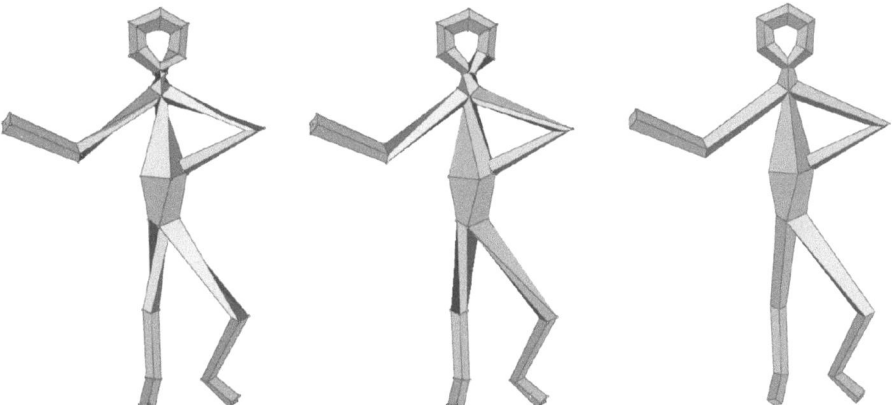

Fig. 2.5 Three meshes based on the same skeleton, the first two exhibit twisting, the last one is the output of our algorithm that minimizes the twist

2.5.1.2 Branches Ending at *Extremities*

For branches ending at extremities, as in, for example, the leg of the stick woman (Figure 2.1), we require a method to distribute the necessary twist along a branch of a skeleton. Such methods appear, for instance, in [20, 21]. We use the algorithm described in [21] to propagate the 0th frame over the branch. The algorithm contains a technique, the so-called double reflection method, which tries to minimize the rotation of the propagated frames along consecutive sampled points of a curve in 3D. In our case the sequence of points is the sequence of vertices of the branch, all joints outside of the end points. For every vertex v_i, $1 \le i \le l$, we define the associated tangent vectors z_i as a unit vector following the direction $((v_i - v_{i-1}) + (v_{i+1} - v_i))$, that is, the average direction between the inward and outward edges of the joint, which is the z-axis of the local frame. Following [21], the local orthonormal frame (x_i, y_i, z_i), associated with the node v_i, is computed by applying two consecutive reflections of the frame $(x_{i-1}, y_{i-1}, z_{i-1})$ of the previous vertex v_{i-1} of the branch. Specifically, $x_i = R_2 R_1 x_{i-1}$, where $R_j = I - 2(n_j n_j^T)/(n_j^T n_j)$, $j = 1, 2$ are the reflections relative to the planes with normals $n_1 = v_i - v_{i-1}$ and $n_2 = (v_i + z_i) - (v_i + R_1 z_{i-1})$, respectively. The vector y_i is then the cross product of the two axes z_i, x_i, i.e., $y_i = z_i \times x_i$.

Next, we propagate the spherical quad from the originating branch node of the branch along the axis z_i and on the plane defined by the other two axes x_i, y_i. Specifically, given that we have a well-oriented frame to the quad of the node v_{i-1}, we define the next (propagated spherical) quad $(q_i^1, q_i^2, q_i^3, q_i^4)$ of the node v_i of radius r_i and with local frame (x_i, y_i, z_i) as

$$q_i^1 = v_i + r_i x_i + r_i y_i$$

$$q_i^2 = v_i + r_i x_i - r_i y_i$$

$$q_i^3 = v_i - r_i x_i - r_i y_i$$

$$q_i^4 = v_i - r_i x_i + r_i y_i$$

2.5.1.3 Branches Ending at Another *Branch Point*

In the setting where the last node of the branch is not an extremity but another *branch point* v_l, we need to match the last generated propagated spherical quad to the associated spherical quad. Since v_l is a branch point, a spherical quad corresponding to the edge (v_{l-1}, v_l) is already defined from the sphere partition at v_l (as detailed in Section 2.4); we denote it $(p^i)_{i=1...4}$. If needed, we orient this quad (coming from the partition at v_l) consistently by minimizing the Euclidean distance between every vertex of the propagated quad $(q_l^1, q_l^2, q_l^3, q_l^4)$ and the assigned spherical quad (p^1, p^2, p^3, p^4) (four possibilities are considered, offsetting mod 4 the four indices of the quad vertices in reverse order). We update the indices of the $(p_i)_{i=1...4}$ to match the indices of the $(q_l^i)_{i=4...1}$.

2.5.1.4 Generating the Scaffold Mesh

For each edge, in each branch, we create four new faces that are associated with the edge (v_{i-1}, v_i). The four scaffold quads are:

$$(q_i^1, q_i^2, q_{i-1}^2, q_{i-1}^1), (q_i^2, q_i^3, q_{i-1}^3, q_{i-1}^2), (q_i^3, q_i^4, q_{i-1}^4, q_{i-1}^3), (q_i^4, q_i^1, q_{i-1}^1, q_{i-1}^4).$$

If the branch ends with an extremity, we generate an additional scaffold quad to close the mesh: $(p_l^4, p_l^3, p_l^2, p_l^1)$.

Algorithm 2.2 Sleeving the branches

Data: A skeleton (G, p, r), an ordering of the vertices given by the depth first search, and a partition of each sphere corresponding to a branch point into spherical quads.

1: **function** CREATESLEEVES(G)
2:     **for** each branch in G given by the sequence $v_0, \ldots, v_k$ of the skeleton vertices **do**
3:         $p_1 \ldots p_4 \leftarrow$ the face in the partition of the sphere at v_0 associated with the neighbor v_1
4:         Create a local orthonormal frame (x_0, y_0, z_0) for the v_0
5:         $q_0^1 \leftarrow p_3$
6:         $q_0^2 \leftarrow p_4$
7:         $q_0^3 \leftarrow p_1$
8:         $q_0^4 \leftarrow p_2$
9:         Align the frame to the spherical quad for the vertex v_0, $\{p_j^1 \ldots p_j^4\}$ if needed
10:         **for** each node $v_i, i > 0 \in s_j$ and corresponding radius r_i **do**
11:             Create the local frame (x_i, y_i, z_i) at v_i, using the 'Double Reflection Method' [21].
12:             $q_i^1 \leftarrow v_i + r_i * x_i + r_i * y_i$
13:             $q_i^2 \leftarrow v_i + r_i * x_i - r_i * y_i$
14:             $q_i^3 \leftarrow v_i - r_i * x_i - r_i * y_i$
15:             $q_i^4 \leftarrow v_i - r_i * x_i + r_i * y_i$
16:             **if** $i = k$ and v_k is a branch node **then**
17:                 Align the propagated spherical quad associated with node v_k to the frame.
18:             **end if**
19:             Add to the Mesh the four scaffold quad faces $f_1, \ldots, f_4$:
20:             $f_1 \leftarrow (q_i^1, q_i^2, q_{i-1}^2, q_{i-1}^1)$
21:             $f_2 \leftarrow (q_i^2, q_i^3, q_{i-1}^3, q_{i-1}^2)$
22:             $f_3 \leftarrow (q_i^3, q_i^4, q_{i-1}^4, q_{i-1}^3)$
23:             $f_4 \leftarrow (q_i^4, q_i^1, q_{i-1}^1, q_{i-1}^4)$
24:             **if** $i = k$ and v_k is an extremity **then**
25:                 Add to the Mesh face $f_5 = (q_i^4, q_i^3, q_i^2, q_i^1)$
26:             **end if**
27:         **end for**
28:     **end for**
29:     **return** Coarse quad-mesh surrounding the skeleton
30: **end function**

2.6   Implementation and Results

The implementation was written in MATLAB and C++, with examples generated in Rhino3D/Grasshopper3D. One consideration in viewing these results is that the scaffold is intended as a base mesh and is useful for its topology and orientation but is not a final mesh representing the geometry of intended shape sketched by the skeleton. The visual appearance of the result is thus not the emphasis of the current project. The emphasis is on the coarseness of the generated mesh that is free of any T-junctions: the scaffold is obtained by connecting the spherical quads at each end of a branch (i.e., between two branch points or a branch point and an extremity). We illustrate in this section the effectiveness of our algorithm on several examples.

Figure 2.6 shows a spiral starting at a degree 3 node and illustrates the rotation-minimizing section along the long branch.

The second example consists of a simple tree skeleton, in two cases. First, we consider nonintersecting spheres with constant radius at each branch point of the skeleton. The skeleton and resulting scaffold are shown in Figure 2.7. We then assign to the branch point and joints a larger radius leading to intersecting spheres (Figure 2.8b). The output is shown in Figure 2.8, and we note that the mesh generated is not self-intersecting.

Figure 2.9 is the quad scaffold of a skeleton with branch points of degree 4 and 3 (and varying radii) and a graph with a cycle. The "Stick Woman" illustrates the simplicity of the generated mesh, as every edge of the graph generates only four quads on the scaffold; an additional quad at each extremity is created to make a closed manifold mesh.

Fig. 2.6 A simple skeleton with constant radii and 2- and 3-valent vertices. We can see in the long spiral the consistent rotation of the frame along the turning sleeve

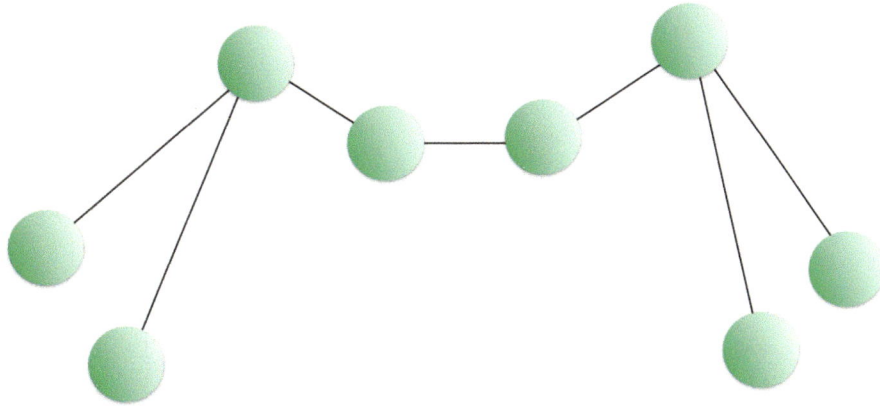

(a) A skeleton with equal radii at each vertex

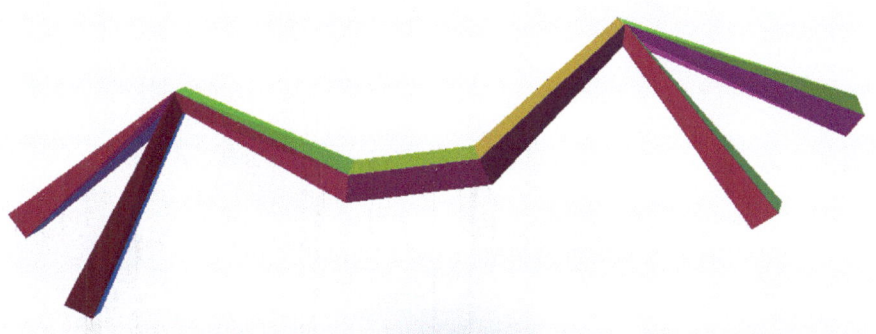

(b) The uniform quad scaffold mesh around the skeleton

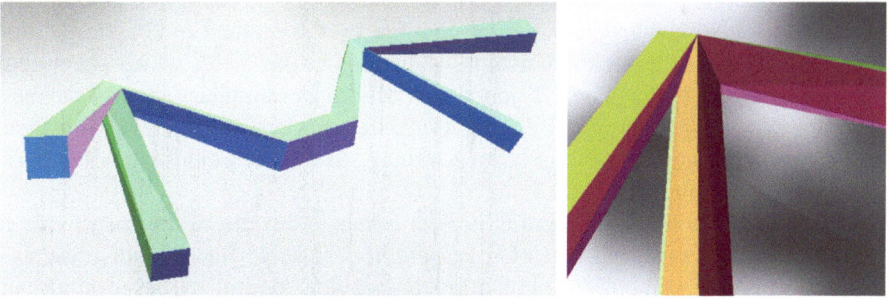

(c) The scaffold mesh from a different viewpoint   (d) Around a *branch point*

Fig. 2.7 A scaffold simple skeleton with unit radii and 2- and 3-valent vertices. Note the regularity and stitching of the mesh faces around branch points (**d**)

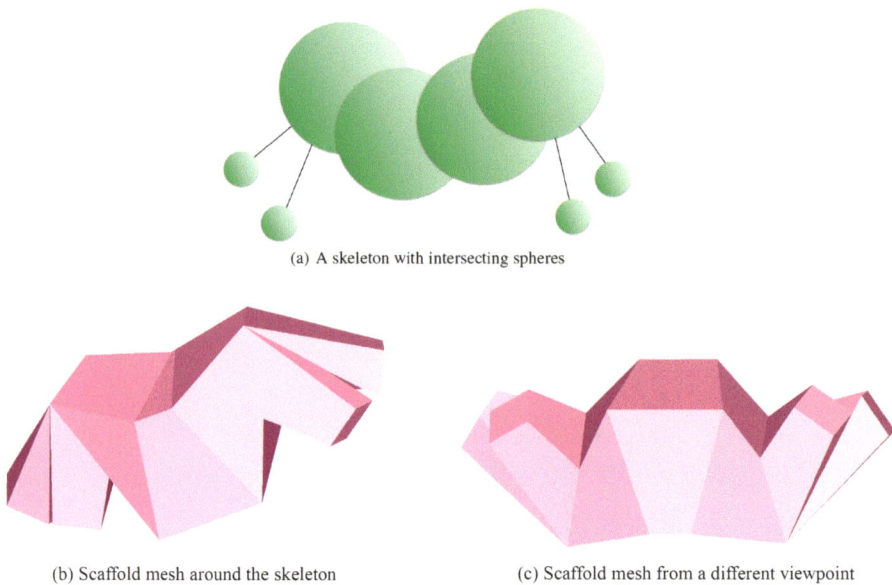

(a) A skeleton with intersecting spheres

(b) Scaffold mesh around the skeleton   (c) Scaffold mesh from a different viewpoint

Fig. 2.8 The simple skeleton of Figure 2.7 with varying radii. The large radii lead to intersecting spheres at the interior vertices, but the faces remain quads

2.7   Conclusion, Limitations, and Future Work

In conclusion, we have provided an algorithm for generating a coarse mesh around a skeleton. The constructed scaffold is a first step to skinning the skeleton with an explicit representation and offers several advantages. In particular, we guarantee an exclusively quad mesh. This was not possible by the method of [2]. In addition, by construction we completely avoid T-junctions in the mesh, which was one of the goals of [19]. In that paper, T-junctions occur in a preliminary stage of the mesh creation and are then removed by a rather involved process; moreover, some difficult cases remain. It would therefore be an advantage to use our proposed scaffold as an initial step for both work.

One of the motivations for this project comes from the desire to provide an explicit surface representation of a *convolution surface*. Recent work has used techniques from symbolic computation to provide simplified descriptions and efficient computation of a convolution surface based on a skeleton [9]. However, although these methods give very satisfying smooth surfaces, they lack an explicit (e.g., mesh-based) representation of the resulting surfaces and are therefore computationally expensive to render. Typically a method such as the marching cubes algorithm is required to render the surface. The scaffold mesh constructed in the present work could be used to map a set of representative points from the convolution surface, which would lead to generating a mesh, that is, an explicit representation of the surface.

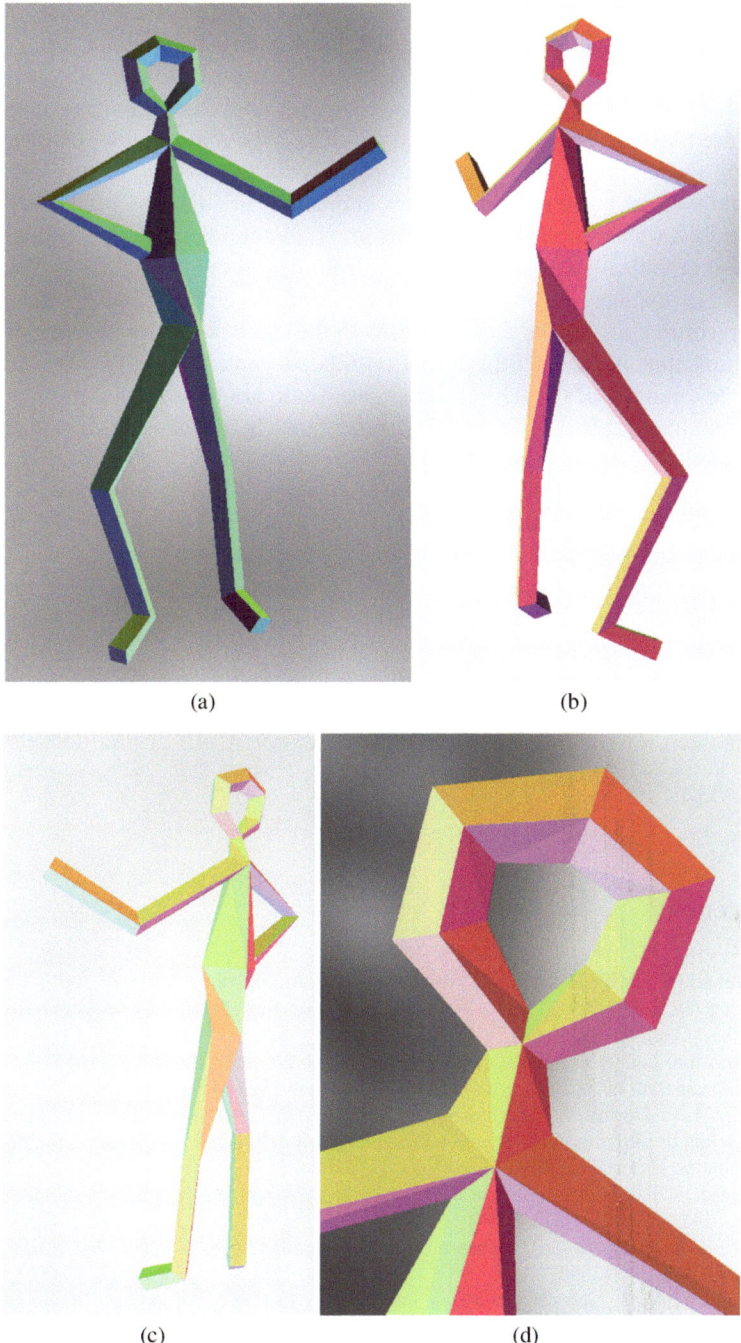

(a)   (b)

(c)   (d)

Fig. 2.9 The scaffold mesh created from the skeleton in Figure 2.1; all faces of the mesh are quads, as well as the section of any sleeve

Several aspects of the present work remain open for investigation. While the sphere partition algorithm of Section 2.4 guarantees a partition of the sphere into spherical quads, that partition is not necessarily unique. In particular, the partition depends on the order in which the branches are considered. Further investigation is required to determine what methods may be available to either standardize this partition (by sliding the vertices appropriately on the surface of the sphere) or to potentially allow some element of user interaction to guide the partition. This may be desirable, for example, in a computer graphics application, where a skeleton may represent an object such as a hand or a creature, and one particular partition of the sphere may enhance its appearance. If the skeleton changes, but not its topology, our algorithm always produces a mesh with the same number of vertices and faces, something that is useful for simulation-related applications. On the other hand, when the topology of the skeleton changes, we do not have any intuition about how the mesh changes. It is an interesting direction to investigate for animation-related applications.

Another consideration for further work is the rigorous treatment of the case of intersecting spheres at the nodes of the skeleton. Finally, the technique that provides an initial frame for the spherical quads needs further research to guarantee that the resulting frame-quad combination is always well aligned. For that reason, a relaxation of the positions of the nodes could be used as a post-processing to improve the result.

Acknowledgements This work has been partly supported by the NSF DMS-1619759 grant and the French Research National Agency (ANR) program CIMI, ANR-11-LABX-0040-CIMI and the German Research Foundation (DFG) within the priority program SPP 1962 under contract number Schu804/15-1.

References

1. Adhikari, M.R.: Basic Algebraic Topology and Its Applications. Springer, Berlin (2016)
2. Bærentzen, J.A., Misztal, M.K., Welnicka, K.: Converting skeletal structures to quad dominant meshes. Comput. Graph. **36**(5), 555–561 (2012)
3. Bærentzen, J.A., Abdrashitov, R., Singh, K.: Interactive shape modeling using a skeleton-mesh co-representation. ACM Trans. Graph. **33**(4), Article 132 (2014)
4. Cani, M.-P., Hornus, S.: Subdivision curve primitives: a new solution for interactive implicit modeling. In: International Conference on Shape Modeling and Applications, pp. 82–88. IEEE Computer Society Press, Washington, D.C. (2001)
5. Farouki, R.T.: Rational rotation-minimizing frames – recent advances and open problems. Appl. Math. Comput. **272**, 80–91 (2016)
6. Goh, W.B.: Strategies for shape matching using skeletons. Comput. Vis. Image Underst. **110**(3), 326–345 (2008)
7. Hitchman, M.P.: Geometry with an Introduction to Cosmic Topology. Jones and Bartlett Publishers, Burlington, MA (2009)
8. Hornus, S., Angelidis, A., Cani, M.-P.: Implicit modelling using subdivision curves. Vis. Comput. **19**(2–3), 94–104 (2003)

9. Hubert, E., Cani, M.-P.: Convolution surfaces based on polygonal curve skeletons. J. Symb. Comput. **47**(6), 680–699 (2012)
10. Igarashi, T., Matsuoka, S., Tanaka, H.: Teddy: a sketching interface for 3D freeform design. In: Proceedings of the 26th Annual Conference on Computer Graphics and Interactive Techniques, pp. 409–416 (1999)
11. Ji, Z., Liu, L., Wang, Y.: B-mesh: a Modeling system for base meshes of 3D articulated shapes. Comput. Graph. Forum **29**(7), 2169–2177 (2010)
12. Jin, X., Tai, C.-L., Feng, J., Peng, Q.: Convolution surfaces for line skeletons with polynomial weight distributions. ACM J. Graph. Tools **6**(3), 17–28 (2001)
13. Lakatos, I.: Proofs and Refutations. Cambridge University Press, Cambridge (1976)
14. Raptis, M., Kirovski, D., Hoppe, H.: Real-time classification of dance gestures from skeleton animation. In: Proceedings of the 2011 ACM SIGGRAPH/Eurographics Symposium on Computer Animation, pp. 147–156 (2011)
15. Srinivasan, V., Akleman, E., Chen, J.: Interactive construction of multi-segment curved handles. In: Proceedings of Pacific Graphics, Beijing, China (2002)
16. Srinivasan, V., Mandal, E., Akleman, E.: Solidifying wireframes. In: Bridges: Mathematical Connections in Art, Music, and Science 2004, Banff, Alberta, Canada (2005)
17. Sundar, H., Silver, D., Gagvani, N., Dickinson, S.: Skeleton based shape matching and retrieval. In: International Shape Modeling, pp. 130–139 (2003)
18. Tagliasacchi, A., Delame, T., Spagnuolo, M., Amenta, N., Telea, A.: 3D skeletons: a state-of-the-art report. Comput. Graph. Forum **35**, 573–597 (2016)
19. Usai, F., Livesu, M., Puppo, E., Tarini, M., Scateni, R.: Extraction of the quad layout of a triangle mesh guided by its curve skeleton. ACM Trans. Graph. **35**(1), Article 6 (2015)
20. Wang, W., Joe, B.: Robust Computation of the rotation minimizing frame for sweep surface modeling. Comput. Aided Des. **29**(5), 379–391 (1997)
21. Wang, W., Juttler, B., Zheng, D., Liu, Y.: Computation of rotation minimizing frames. ACM Trans. Graph. **27**(1), 1–18, Article 2 (2008)

Chapter 3
Convolution Surfaces with Varying Radius: Formulae for Skeletons Made of Arcs of Circles and Line Segments

Alvaro Javier Fuentes Suárez and Evelyne Hubert

Abstract In skeleton-based geometric modeling, convolution is an established technique: smooth surfaces around a skeleton made of curves are given as the level set of a convolution field. Varying the radius or making the surface scale sensitive along the skeleton are desirable features. This article provides the related necessary closed-form formulae of the convolution fields when the skeleton is made of arcs of circle and line segments. For the family of power inverse kernels, closed-form formulae are exhibited in terms of recurrence relationships. These are obtained by creative telescoping. This novel technique is described from a practitioner point of view so as to be applied to other families of kernels or skeleton primitives. The newly obtained formulae are applied to obtain convolution surfaces around $\mathcal{G}^1$ skeleton curves, some of them closed curves. Having arcs of circles in addition to line segments allows to demonstrably improve the visual quality of the generated surface with a lower number of skeleton primitives.

3.1 Introduction

Shape representation based on skeletons has a major role in geometric modeling and animation in computer graphics [10]. Skeletons of 3D shapes are made of curves and surfaces, with a preference for the former. Several techniques are used to create a surface enclosing a volume around a skeleton. In this paper we focus on *convolution surfaces* [5]. This technique provides surfaces with good mathematical features stemming from their definition. It is, for instance, used in interactive modeling environments [2, 41] and sketch-based modeling [1, 39] and applied to model natural shapes [16, 40].

Convolution surfaces are the level sets of a convolution function that results from integrating a *kernel* function K along the *skeleton* S. The mathematical smoothness

A. J. Fuentes Suárez (✉) · E. Hubert
INRIA Méditerranée, BP 93, Sophia Antipolis, France
e-mail: alvaro.fuentes-suarez@inria.fr; evelyne.hubert@inria.fr

© The Author(s) and the Association for Women in Mathematics 2018
A. Genctav et al. (eds.), *Research in Shape Analysis*, Association for Women in Mathematics Series 12, https://doi.org/10.1007/978-3-319-77066-6_3

of the surface obtained depends only on the smoothness of the kernel. In practice, a kernel function (power inverse, Cauchy, compact support, etc.) is selected so as to have closed-form expressions for the convolution functions associated to basic skeleton elements (line segments, triangles, etc.). Since the additivity property of integration makes the convolution function independent of the partition of the skeleton, skeletons are partitioned and approximated by a set of basic elements. The convolution function for the whole skeleton is obtained by adding the convolution functions of the constitutive basic elements. See, for instance, [5, 9, 17, 20–22, 32, 33, 36, 38].

Line segments are the most commonly used 1D basic skeleton elements. When a skeleton consists of curves with high curvature or torsion, its approximation might require a great number of line segments for the convolution surface to look as intended. In this paper we take the stance that arcs of circles form a very interesting class of basic skeleton elements in the context of convolution. This was already argued in [21] for planar skeleton curves. But this can be made more general. Indeed any space curve can be approximated by circular splines in a $\mathcal{G}^1$ fashion [28, 34]. A lower number of basic skeleton elements are then needed to obtain an appealing convolution surface, resulting into better visual quality at lower computational cost.

To model a wider variety of shapes, it is necessary to vary the thickness around the skeleton. Several approaches have been suggested, weighted skeletons [19, 20, 22], varying radius [17], and scale invariant integral surfaces [38], the latter two actually providing a more intrinsic formulation. While general closed-form formulae were obtained for weighted line segments in [19], there has been a lack of generality in terms of closed-form formulae for convolution with varying radius, or scale, over line segments and, even more so, over arcs of circles. This paper addresses this very issue for the family of power inverse kernels, of even degree. For a family of compact support kernels, the formulae are analogous for the convolution of line segments but are substantially more complicated in the convolution of arcs of circle. We thus chose not to present them explicitly in this article. Yet these formulae can be obtained with the advanced computer algebra technique that we present in this paper and the power of which we illustrate on power inverse kernels.

As initiated in [18, 19], the generality for closed-form formulae of the convolution function associated to line segments and arcs of circle for a varying radius or scale is offered in terms of recurrence formulae. Closed-form formulae allow efficient evaluation. With recurrence formulae we elegantly reach a higher level of generality: with a simple code, we can use kernels of any degree. As detailed and illustrated in [19], the code can be automatically generated and optimized within a computer algebra software like Maple. This contrasts with previous works where the formulae for the convolution functions had to be implemented individually for kernels of every degree. The recurrence formulae in [18, 19] mostly drew on integral functions that appear in some classical tables. An innovative approach is taken in this paper. The recurrence formulae we present in this paper are obtained through creative telescoping, an active topic of research in computer algebra. A limited set of pointers is [7, 13, 25] as they provide the background to available software. We introduce the subject and show how this technique can be applied in the context of convolution surfaces.

In Section 3.2, we provide the definition of convolution surfaces associated to a skeleton made of curves, with varying thickness. In Section 3.3 we introduce creative telescoping and describe how it can be used to obtain recurrence formulae for convolution fields. In Section 3.4 we examine the convolution of line segments with varying thickness, providing the closed-form formulae for the convolution functions. In Section 3.5 we turn to arcs of circles. In Section 3.6 we demonstrate the interest of having arcs of circles as basic skeleton elements for convolution and discuss future directions.

3.2   Convolution Surfaces

In this section we recall the basics of convolution surfaces. We first discuss families of kernels that arose in the computer graphics literature. Thereafter, we shall mainly focus on the family of *power inverse* kernels. We then define the convolution function generated by a bounded regular curve and provide alternative definitions of convolution that allow the modeling of shapes with varying thickness around the skeleton.

3.2.1   Kernels

The kernels in use in the literature are given by functions $K : \mathbb{R}^+ \to \mathbb{R}^+$ that are at least continuously differentiable. The argument is the distance between a point in space and a point on the skeleton. Those kernels are decreasing functions on $\mathbb{R}^+$ and strictly decreasing when nonzero $K(r) > 0 \Rightarrow K'(r) < 0$.

The first convolution surfaces [5] were based on the Gaussian kernel $r \mapsto e^{-s\,r^2}$ (also in [4]) that depends on a parameter $s > 0$. The difficulty in evaluating the resulting convolutions prompted the introduction of kernels that provided closed-form expressions for the convolution functions associated to basic skeleton elements. [32, 33] promoted the Cauchy kernel $r \mapsto \frac{1}{(1+\sigma\,r^2)^2}$ after [35] introduced the inverse function $r \mapsto \frac{1}{r}$. For faster convolution [9, 17] introduced the power inverse cube kernel $r \mapsto \frac{1}{r^3}$. [20] also exhibited the benefit of using the quintic inverse $r \mapsto \frac{1}{r^5}$ (Figure 3.1). With the power inverse kernels and the Cauchy kernels, the whole skeleton influences the convolution function at a point, even if infinitesimally. In order to limit the zone of influence of each point of the skeleton, compact support kernels were introduced, mostly as piecewise polynomial functions. To some extent, they allow to limit the unwanted bulges or blending (Figure 3.3). Their use nonetheless necessitates to determine the geometry of the intersection of the skeleton with spheres.

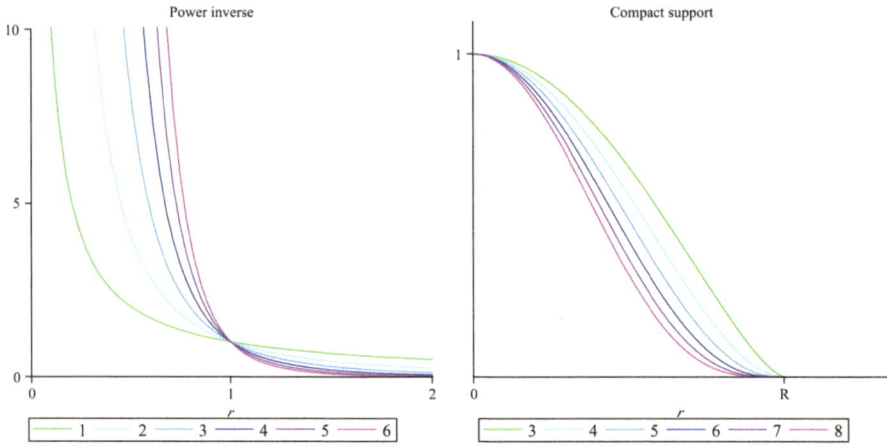

Fig. 3.1 The graphs of the kernel functions p^i and k_R^i, varying i

The family of compact support kernels used in [18] is indexed by $i \in \mathbb{N} \setminus \{0\}$ and given by

$$k_R^i : r \mapsto \begin{cases} \left(1 - \left(\frac{r}{R}\right)^2\right)^{\frac{i}{2}} & \text{if } r < R \\ 0 & \text{otherwise.} \end{cases}$$

To obtain convolution surfaces that are at least continuously differentiable, only k_R^i for $i \geq 3$ should be considered. For $i < 3$, k_R^i is not differentiable at $r = R$. The case $i = 4$ is actually the case considered in [23, 33]. As we increase i though, we obtain smoother shapes.

This paper is mostly concerned with the family of power inverse kernels though. They are indexed by $i \in \mathbb{N} \setminus \{0\}$ and given by

$$p^i : r \mapsto \left(\frac{1}{r}\right)^i$$

The convolution surfaces obtained with a power inverse kernel always enclose the skeleton since these functions tend to infinity when approaching the skeleton. As exploited in [18, 19], the closed-form formulae of convolution functions using the family of Cauchy kernels $c_\sigma^i : r \mapsto \left(\frac{1}{1+\sigma r^2}\right)^{\frac{i}{2}}$ differ only slightly from the ones using power inverse kernels.

3.2.2 Notations

Typically $P = (p_1, p_2, p_3)^T \in \mathbb{R}^n$ represents a point in space, and $A = (a_1, a_2, a_3)^T$ and $B = (b_1, b_2, b_3)^T$ represent the end points of a line segment $[AB]$ or arc of circle $\overset{\frown}{A^O B}$ with center O. The straight line through A and B is denoted as (AB). Then $\overrightarrow{AP}$ represent the vector from A to P. In the following $\overrightarrow{u} = (u_1, u_2, u_3)^T$, $\overrightarrow{v} = (v_1, v_2, v_3)^T \in \mathbb{R}^n$ also represent vectors. The scalar product of two vectors $\overrightarrow{u} = (u_1, u_2, u_3)^T$ and $\overrightarrow{v} = (v_1, v_2, v_3) \in \mathbb{R}^n$ is then $\overrightarrow{u} \cdot \overrightarrow{v} = u_1 v_1 + u_2 v_2 + u_3 v_3$. The distance between two points A and B is denoted as $|AB| = \sqrt{(b_1 - a_1)^2 + (b_2 - a_2)^2 + (b_3 - a_3)^2}$

3.2.3 Convolution of Regular Curves

We assume here that the *skeleton* can be partitioned into *regular patches* defined by bounded regular curves parametrized by an interval $[a, b]$ of $\mathbb{R}$. The convolution function for the skeleton is then obtained by summing the convolution functions for each of those patches.

Consider a parametrized curve $\Gamma : [a, b] \subset \mathbb{R} \to \mathbb{R}^3$. It is a regular curve if Γ is continuously differentiable and Γ' does not vanish. The infinitesimal arclength is then $|\Gamma'(t)| \, dt$. The convolution function based on $\mathcal{S} = \Gamma([a, b])$ at a point $P \in \mathbb{R}^3$ is then defined by

$$\mathcal{C}_\Gamma^K(P) = \int_a^b K\left(|P\Gamma(t)|\right) |\Gamma'(t)| \mathrm{d}t.$$

The integral is independent of the (regular) parametrization of the curve used. Convolution functions with a power inverse kernel $K = p^i$ are infinitely differentiable outside of the curve $\Gamma([a, b])$. Convolution functions with a compact support kernel $K = k_R^i$ are $\lfloor \frac{i-1}{2} \rfloor$ continuously differentiable.

As the inverse image of a closed set by a k-continuously differentiable map, $k \geq 1$, the resulting *convolution surfaces* $\{P \in \mathbb{R}^3 \,|\, \mathcal{C}_\Gamma^K(P) = \kappa\}$ are closed (in a topological sense) and smooth, provided κ is not a critical value[1] of $\mathcal{C}_\Gamma^K$. It is the boundary of a smooth *three*-dimensional manifold $V_\kappa = \{P \in \mathbb{R}^3 \,|\, \mathcal{C}_\Gamma^K(P) \geq \kappa\}$. Furthermore V_κ and $V_{\kappa'}$ are diffeomorphic provided that there are no critical values in the interval $[\kappa, \kappa']$ [27, Theorem 3.1]. With the power inverse kernel, the skeleton is in V_κ for any $\kappa > 0$. This is not always the case with a Cauchy or compact support kernel; V_κ can even be the empty set for too high values of κ.

---

[1]A critical point of a function $f : (x, y, z) \mapsto f(x, y, z)$ is a point (x_0, y_0, z_0) at which the gradient (f_x, f_y, f_z) of f vanishes. A critical value of f is the value $f(x_0, y_0, z_0)$ of f at a critical point (x_0, y_0, z_0).

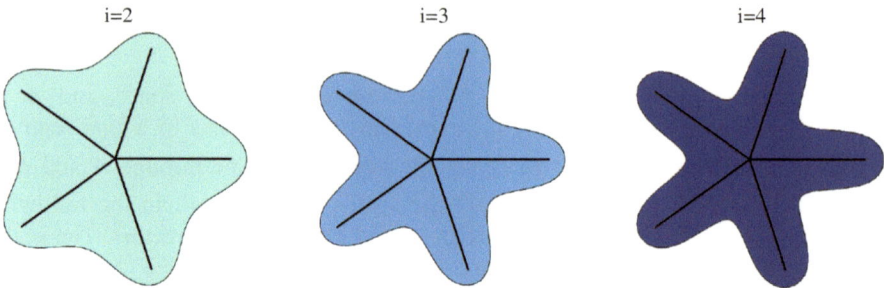

Fig. 3.2 Convolution curves for a set of segments with power inverse kernel p^2, p^3, p^4. The convolution function for the whole set is obtained as the sum of the convolution functions for each line segment. The level set is adapted to have identical thickness at the tips. Note that sharpness increases from left to right

Fig. 3.3 Convolution curves based on two parallel segments. The top line uses the cubic inverse kernel p^3 and the bottom line a compact support kernel k_R^3. Columns correspond to an identical sought thickness. For the cubic inverse kernel, a bulge and then a blend appear between the line segments, while the convolution surfaces get only thicker with the compact support kernel

3.2.4 Discussion on the Choice of a Kernel

With compact support kernels k_R^i, the smoothness of the convolution surface increases with i. With power inverse kernels p^i the convolution functions are smooth at all points outside the skeleton. Yet, as i increases, the convolution surface is *sharper* around the skeleton. This is illustrated in Figure 3.2.

When the convolution function has a critical point, chances are that there is a change in the topology of the convolution surface as it goes through the critical value [27]. This is illustrated in Figure 3.3 with a skeleton made of two line segments. The convolution function has a critical and the convolution surface through this point has a singularity. The corresponding level set is a transition from bulging to blending, from two connected components to a single component. Figure 3.3 also illustrates the fact that compact support kernels allow to dismiss the influence of skeleton elements that are at distance more than R, thus avoiding some of the bulging and blending that appear for the kernels with infinite support kernel.

3.2.5 Varying Thickness

Several alternatives have been introduced for varying the thickness of the convolution surface along the skeleton. A first idea was to use a *weight* function along the skeleton [5]. For a skeleton given by a regular curve $\Gamma : [a, b] \to \mathbb{R}^3$, one uses a weight function $w : [a, b] \to \mathbb{R}$. The convolution function is then defined as

$$\mathcal{C}_{\Gamma,w}^{K}(P) = \int_a^b w(t) K \left(|P\Gamma(t)| \right) |\Gamma'(t)| \, \mathrm{d}t.$$

The convolution function is now dependent on the parametrization used for the skeleton curve.

Polynomial weights were studied in [19–22], where closed-form formulae were obtained for particular cases of weighted convolution. The drawback of this approach was illustrated in [19, Figure 9]: the influence of the weight diminishes as the degree of the kernel increases. Alternative more intrinsic formulations were proposed in [17, 38].

3.2.5.1 Varying Radius as Proposed by Hornus *et al.* [17]

In [17] Hornus *et al.* proposed a different way of computing the convolution function. This allows the user to modify the shape of the final surface according to a radius assigned to each point in the skeleton: the distance is divided by the radius at the corresponding point in the skeleton. For a skeleton curve $\Gamma : [a, b] \to \mathbb{R}^3$, the radius is given by a function $\rho : [a, b] \to \mathbb{R}^+$. The convolution function is then defined by

$$\mathcal{H}_{\Gamma,\rho}^{K}(P) = \int_a^b K \left(\frac{|P\Gamma(t)|}{\rho(t)} \right) |\Gamma'(t)| \, \mathrm{d}t.$$

If additional precautions are not, imposed on the radius function ρ, this convolution function depends on the parametrization of the curve Γ. In practice it makes sense to have a radius function that is approximately linear in the arclength.

3.2.5.2 SCALIS as Proposed by Zanni *et al.* [38]

An alternative convolution introduced in [38] allows to properly model shapes with pieces at different scales. This lead to the name SCALe invariant Integral Surfaces (SCALIS). For a regular curve $\Gamma : [a, b] \subset \mathbb{R} \to \mathbb{R}^3$, the convolution function is

$$\mathcal{S}^K_{\Gamma,\lambda}(P) = \int_a^b K\left(\frac{|P\Gamma(t)|}{\lambda(t)}\right) \frac{|\Gamma'(t)|}{\lambda(t)} \, dt,$$

where $\lambda : [a, b] \to \mathbb{R}^+$ is called the *scale function*. Notice that λ plays a similar role as ρ in the formulation by Hornus. To apprehend all the good features of this new definition of convolution surface, the reader is referred to [36, 38]. Let us just observe the case where λ is a constant. If we write $\lambda \cdot P$ and $\lambda \cdot \Gamma$ for the point and the regular curve obtained through a homothety (*a.k.a* homogeneous dilatation or scaling) with ratio λ, then

$$\mathcal{S}^K_{\lambda\cdot\Gamma,\lambda}(\lambda \cdot P) = \mathcal{C}^K_\Gamma(P) \quad \text{or equivalently} \quad \mathcal{S}^K_{\Gamma,\lambda}(P) = \mathcal{C}^K_{\lambda^{-1}\cdot\Gamma}\left(\lambda^{-1} \cdot P\right).$$

This implies that the convolution surface of equation $\mathcal{S}^K_{\lambda\cdot\Gamma,\lambda}(P) = c$ is homothetic to the convolution surface $\mathcal{C}^K_\Gamma(P) = c$ with a ratio λ.

3.3 Determining General Closed-Form Formulae Through Recurrence

Closed-form formulae for convolution fields have been studied for several skeleton primitives and kernels [18–22, 33, 36, 37]. Given an appropriate parametrization of a skeleton primitive and a kernel, the closed-form formulae for the convolution fields are very often obtained thanks to symbolic integration [8], which is implemented in computer algebra software as Maple or Mathematica. The field function for the pair consisting of the skeleton primitive and kernel can then be implemented with a view on optimizing its evaluation cost. As different kernels change the properties of the convolution surface, even if lightly, it is interesting to offer, in the same geometric modeling software, alternative kernels for convolution. The same is true for skeleton primitives. As different kernels may bring related closed-form formulae, one seeks to optimize the code by taking advantage of the common subexpressions [22, Equations 11–14]. A greater level of generality was offered with the new approach in [18, 19] where common subexpressions were encapsulated into low order recurrence formulae: kernels were grouped into families, each indexed by an integer, and, for each family, recurrence relationships were exhibited for the convolution function field, indexed by the same integer. The recurrence relationships that appeared in [18, 19] mostly came from tables of known integrals, which made this strategy uneasy to generalize.

We present here a technique that allows to obtain algorithmically the recurrence formulae on the convolution fields for families of kernels and a given skeleton primitive. This unified approach relies on creative telescoping, and we think it is a valuable contribution of ours of realizing how to apply this technique in the context of convolution surfaces. Within computer algebra software, recurrence relationships can be translated into optimized C code, as illustrated in [19]. But this approach should be balanced against the use of highly accurate numerical integration routines based on quadratures [31]. Indeed, some closed-form formulae might be judged too complex to result in efficient code for evaluating the convolution functions.

In this section we give a brief introduction to *Creative Telescoping* (CT) and illustrate how to use some specialized software for our purpose. In Sections 3.4 and 3.5, we show some specific recurrences we obtained with CT for line segments and arcs of circle with power inverse and compact support kernels. All the recurrences we provide in this article, as well as the recurrence equations, can be checked by a simple differentiation.

3.3.1 Creative Telescoping

We introduce here creative telescoping. This is an active research field [12], with new algorithms and new applications appearing every year. For a complete (but still gentle) introduction, we refer the reader to [25] and to [14] for a formal development of the ideas we present here.

Creative telescoping comes in several guises depending on the application. We want to find recurrence relations for the integrals in the convolution field. For this case CT works with *differential* (D_x) and *shift* (S_n) operators. These operators act by differentiating or incrementing by one (in the respective variable) the input expression f, i.e., $D_x f(x, n) = \frac{\partial}{\partial x} f(x, n)$ and $S_n f(x, n) = f(x, n + 1)$. Here x and n stand for continuous and discrete variables, respectively.

Higher-order linear operators are constructed as elements of the vector space $\mathbb{O}$ spanned by the symbols of the form $D_x^{\alpha_1} D_y^{\alpha_2} \ldots S_n^{\beta_1} S_m^{\beta_2} \ldots$ with coefficients in $\mathbb{F} = \mathbb{K}(x, y, \ldots, n, m, \ldots)$, the field of rational functions over a field $\mathbb{K}$ (of characteristic 0). $\mathbb{O}$ is actually a $\mathbb{F}$-algebra $\mathbb{F}\langle D_x, D_y, \ldots, S_n, S_m \rangle$ in which the generators $(D_x, D_y, \ldots, S_n, S_m)$ commute pairwise (for instance, $D_x S_n = S_n D_x$), but the commutation of an operator with an element in the field is not trivial:

$$D_x p = p D_x + \frac{\partial}{\partial x} p \quad \text{and} \quad S_n p = p|_{n \to n+1} S_n.$$

$\mathbb{O}$ is a so called Ore algebra. Ore algebras are defined for operators that generalize both derivations and shifts [29]. In this context one introduces the concept of ∂-finite functions [14]. These are often called *holonomic* functions though there are subtleties between the two notions. Creative telescoping takes ∂-finite functions as input. Basically, a ∂-finite function f is uniquely and well defined as the zero of a set of linear operators in $\mathbb{O}$ with prescribed values for a finite subset of its derivatives or shifts (the *initial conditions*). It is then the case that for any operator ∂ (of the form D_x or S_n), there is an integer m such that $\{\partial^k f \mid k = 0..m\}$ is linearly dependent over $\mathbb{F}$.

For illustration consider

$$f(t, i, k) = \frac{(\lambda + \delta t)^k}{(at^2 - 2bt + c)^i}, \text{ with } a, b, c, \delta, \lambda \in \mathbb{R}.$$

For the operators D_t, S_i, and S_k with $\mathbb{F} = \mathbb{R}(t, i, k)$, we see that f is ∂-finite by observing that

$$S_k f = \frac{(\lambda + \delta t)^{k+1}}{(at^2 - 2bt + c)^i} = (\lambda + \delta t) f, \quad S_i f = \frac{(\lambda + \delta t)^k}{(at^2 - 2bt + c)^{i+1}} = \frac{1}{at^2 - 2bt + c} f,$$

and

$$D_t f = \left(\frac{\delta k}{\lambda + \delta t} - 2i \frac{at - b}{at^2 - 2bt + c} \right) f.$$

Given a complete set of operators annihilating the function, the purpose of creative telescoping is to produce operators $P \in \mathbb{O}$ such that

$$P = \mathcal{L} - D_t C \quad \text{and} \quad Pf = 0,$$

with $\mathcal{L}, C \in \mathbb{O}$ and where $\mathcal{L}$ depends neither on D_t nor on t. That is, $\mathcal{L}$ is a linear combination of $S_i^{\beta_1} D_k^{\beta_2}$ over $\mathbb{K}(i, k)$. Because of its total independence on t, $\mathcal{L}$ commutes with an integration operator with respect to t so that

$$0 = \int_{t_0}^{t_1} Pf \, dt = \int_{t_0}^{t_1} \mathcal{L} f \, dt - \int_{t_0}^{t_1} D_t C f \, dt = \mathcal{L} \left(\int_{t_0}^{t_1} f \, dt \right) - [Cf]_{t=t_0}^{t=t_1}.$$

Thus $F(i, k) = \int_{t_0}^{t_1} f(t, i, k) dt$ satisfies the recurrence $\mathcal{L} F(i, k) = g(t_1) - g(t_0)$, where $g = Cf$. The operator $\mathcal{L}$ is called the *telescoper* and C the *certificate*. We thus have obtained a recurrence relationship on the functions $F(i, k) = \int_{t_0}^{t_1} f(t, i, k) \, dt$. For a given skeleton primitive and a family of kernels, the convolution fields are functions of this type. Creative telescoping thus provides a general unified approach to obtain the recurrence formulae on these.

Note that creative telescoping refers to a panel of algorithms. There is no canonical output. For instance, one might trade a telescoper for a telescoper with lower degree coefficient [11] or a telescoper computed with a more efficient heuristic [24]. Furthermore, specialized lower complexity algorithms exist for subclasses of functions, like rational, hypergeometric, or hyperexponential functions [6, 7]. The function f we started with is actually *hypergeometric* since

$$\frac{D_t f}{f} \in \mathbb{F}, \quad \frac{S_i f}{f} \in \mathbb{F}, \quad \text{and} \quad \frac{S_k f}{f} \in \mathbb{F}.$$

3.3.2 *Practical Use*

There are several implementations of CT available; a (perhaps limited) set is
`Mgfun`[2] by F. Chyzak (in Maple), `HolonomicFunctions`[3] by C. Koutschan
(in Mathematica), and `MixedCT`[4] by L. Dumont (in Maple). The descriptions of
the algorithms implemented are in the respective publications [7, 14, 24]. Here,
as Maple users, we describe an example of use with `Mgfun`, following up on the
running example of previous paragraph.

The operators D_t, S_i, S_k are expressed in `Mgfun` by the `t::diff`, `i::shift`,
and `k::shift` directives. In order to get recurrence formulae for the integral

$$F(i, k) = \int_{t_0}^{t_1} f(t, i, k) dt = \int_{t_0}^{t_1} \frac{(\lambda + \delta t)^k}{(at^2 - 2bt + c)^i} dt,$$

one simply calls the command `creative_telescoping(f(t,i,k),`
`[k::shift,i::shift], t::diff)`. It is indeed practical that an
appropriate system of linear operators annihilating f is computed internally. The
class of expressions for which this works is given by the closure properties of
∂-finite functions [14] and accounted for in the documentation of the command
`dfinite_expr_to_sys` in the Mgfun package.

The command `creative_telescoping(f(t,i,k), [k::shift,`
`i::shift], t::diff)` outputs two pairs $(\mathcal{L}, C)$. To interpret the output of
`creative_telescoping`, the reader must refer to Mgfun documentation. The
first pair consists of

$$\mathcal{L}_1 = 2i \left(ac - b^2\right) \left(\lambda^2 a + 2\lambda b \delta + \delta^2 c\right) S_i - a (2i - 2 - k)(a\lambda + \delta b) S_k$$

$$+ \left(2b^2 (i - k - 1) - a c (2i - k - 1)\right) \delta^2 - 2ab\lambda (k + 1)\delta - a^2\lambda^2 (k + 1)$$

and

$$C_1 = (\lambda + \delta t) \left(abt - 2b^2 + ac\right) \delta + a\lambda (at - b),$$

while the second pair consists of

$$\mathcal{L}_2 = a(k - 2i + 3) S_k^2 + 2(i - k - 2)(a\lambda + b\delta) S_k + (k + 1)(a\lambda^2 + 2b\delta\lambda + \delta^2 c)$$

---

[2]https://specfun.inria.fr/chyzak/mgfun.html

[3]http://www.risc.jku.at/research/combinat/software/ergosum/RISC/HolonomicFunctions.html

[4]http://mixedct.gforge.inria.fr

and

$$C_2 = -\delta \ (\lambda + \delta\, t) \left(at^2 - 2\,bt + c\right).$$

This latter, for instance, translates to the following recurrence, which will be used in Section 3.4:

$$a\,(k-2i+3)\,F\,(i,k+2)+2(i-k-2)(a\lambda+b\delta)\,F\,(i,k+1)$$

$$+(k+1)(a\lambda^2+2b\delta\lambda + \delta^2 c)\,F\,(i,k)=\left[\frac{-\delta(\delta t+\lambda)^{k+1}}{(at^2-2bt+c)^{i-1}}\right]_{t_0}^{t_1}.$$

Note that this latter telescoper involves S_k and not S_i. The set of pairs $\{(\mathcal{L}_1, C_1), (\mathcal{L}_1, C_2)\}$ is indeed *minimal* in a sense that we do not wish to make precise here but that depends on the order of k and i in the input. To give a sense of this order, observe that the output of `creative_telescoping(f(t,i,k)`, `[i::shift,k::shift]`, `t::diff)` also consists of two pairs $(\mathcal{L}, C)$, but the second one involves solely S_i.

3.4 Convolution with Line Segments

We examine the convolution of line segments for power inverse kernels, with varying radius or scale.[5] First we express the convolution functions, with varying radius or scale, in terms of an integral function indexed by two integers:

$$I_{i,k}(a, b, c, \lambda, \delta) = \int_{-1}^{1} \frac{(\lambda + \delta\, t)^k}{(a\, t^2 - 2\,b\, t + c)^i}\,dt$$

We then provide recurrence formulae on this integral so as to have all the convolution functions for line segments with (even) power inverse kernels. The recurrence relationships we exhibit can be adapted to work for all powers. Furthermore, though we do not give details, these recurrences also allow to deal with the convolution of line segments with the family of compact support kernels, by taking i to be a negative integer. We choose to restrict here to even powers as they provide easier formulae to evaluate (odd power inverse kernels bring out elliptic functions in the convolution of arcs of circles and planar polygons). This does not affect too much the variety of shapes we can obtain.

---

[5]Results for convolution of weighted line segments with power inverse and Cauchy kernels can be found in [19].

3.4.1 Integrals for Convolution

Two points $A, B \in \mathbb{R}^3$ define the line segment $[AB]$. A regular parametrization for this line segment is given by $\Gamma : [-1, 1] \to \mathbb{R}^3$ with $\Gamma(t) = \frac{A+B}{2} + \frac{B-A}{2} t$. Therefore for a point $P \in \mathbb{R}^3$ we have

$$4\,|P\Gamma(t)|^2 = |AB|^2\,t^2 - 2\,\overrightarrow{AB} \cdot \overrightarrow{CP}\,t + |CP|^2 \text{ where } C = \frac{A+B}{2}$$

is the midpoint of the line segment $[AB]$. Hence $|\Gamma'(t)| = \frac{|AB|}{2}$.

The simple convolution of this line segment with the power inverse kernel p^{2i} is thus given by

$$\mathcal{C}^{2i}_{[AB]}(P) = \frac{|AB|}{2} \int_{-1}^{1} \frac{1}{|P\Gamma(t)|^{2i}}\,dt = \frac{|AB|}{2}\,\mathbf{I}_{i,0}\left(\frac{1}{4}|AB|^2, \frac{1}{4}\overrightarrow{AB} \cdot \overrightarrow{CP}, \frac{1}{4}|CP|^2, \lambda, \delta\right).$$

If we choose the radius function $\rho : [a, b] \to \mathbb{R}$ to be linear in the arclength, we can find $\lambda, \delta \in \mathbb{R}$ such that $\rho(t) = \lambda + \delta t$. Convolution with varying radius is then given by

$$\mathcal{H}^{2i}_{[AB],\rho}(P) = \int_{-1}^{1} \frac{(\lambda + \delta t)^{2i}}{|P\Gamma(t)|^{2i}} \frac{|AB|}{2}\,dt$$

$$= \frac{|AB|}{2}\,\mathbf{I}_{i,2i}\left(\frac{1}{4}|AB|^2, \frac{1}{4}\overrightarrow{AB} \cdot \overrightarrow{CP}, \frac{1}{4}|CP|^2, \lambda, \delta\right).$$

If we now take the scale function to be $\Lambda(t) = \lambda + \delta t$, then

$$\mathcal{S}^{2i}_{[AB],\Lambda}(P) = \int_{-1}^{1} \frac{(\lambda + \delta t)^{2i}}{|P\Gamma(t)|^{2i}} \frac{|AB|}{2(\lambda + \delta t)}\,dt$$

$$= \frac{|AB|}{2}\,\mathbf{I}_{i,2i-1}\left(\frac{1}{4}|AB|^2, \frac{1}{4}\overrightarrow{AB} \cdot \overrightarrow{CP}, \frac{1}{4}|CP|^2, \lambda, \delta\right).$$

3.4.2 Closed Forms Through Recurrence Formulae

First of all, given that

$$\mathbf{I}_{1,0}(a, b, c, \lambda, \delta) = \frac{1}{\sqrt{ac - b^2}} \left[\arctan\left(\frac{a\,t - b}{\sqrt{ac - b^2}}\right) \right]_{-1}^{1}$$

we can determine $I_{i,0}(a, b, c, \lambda, \delta)$ for all $i \geq 1$, thanks to the recurrence relationship

$$2i\left(ac - b^2\right) I_{i+1,0} + (1 - 2i)\, a\, I_{i,0} = \left[\frac{at - b}{\left(at^2 - 2bt + c\right)^i}\right]_{-1}^{1}.$$

One then observes that

$$2\, a\, (1 - i)\, I_{i,1} = 2\, (3 - i)\, (a\lambda + b\delta)\, I_{i,0} - \left[\frac{\delta}{\left(at^2 - 2bt + c\right)^{i-1}}\right]_{-1}^{1}$$

This recurrence is actually obtained by specializing the following recurrence to $k = -1$

$$a\, (k - 2\, i + 3)\, I_{i,k+2} + 2\, (i - 2 - k)\, (a\lambda + b\delta)\, I_{i,k+1}$$

$$+ (k+1)\left(a\lambda^2 + c\delta^2 + 2\, b\lambda\,\delta\right) I_{i,k} = \left[\frac{-\delta\,(\lambda + \delta\, t)^{k+1}}{\left(at^2 - 2\, bt + c\right)^{i-1}}\right]_{-1}^{1}$$

One can thus determine $I_{i,k}$ for all $i \geq 1$ and $k \geq 0$ and therefore $I_{i,2i}$ and $I_{i,2i-1}$ that are needed for convolution with varying radius or scale.

Alternatively, to determine the convolution with varying radius, we can consider the recurrence

$$2i\,(i+1)\, a\left(ac - b^2\right) I_{i+2,2i+4}$$

$$-i\left(\lambda\, a\,(1 + 2\, i)\,(2\, b\delta + a\lambda) + \left((4\, i + 5)ca - 2\,(i + 2)b^2\right)\delta^2\right) I_{i+1,2i+2}$$

$$+\delta^2(i + 1)(1 + 2\, i)\left(a\lambda^2 + c\delta^2 + 2\, b\lambda\,\delta\right) I_{i,2i} = \left[\frac{(\lambda + \delta\, t)^{2i+1}}{\left(at^2 - 2\, bt + c\right)^{i+1}}\, C\right]_{-1}^{1}$$

where

$$\begin{aligned}
C = &\left((ac + 2\, b^2 i)\, t^2 - bc\,(3\, i + 2)\, t + c^2\,(i + 1)\right)\delta^3 \\
&+ \left(2\, ab\,(1 + 2\, i)\, t^2 - \left(2\,(i + 2)\, b^2 + 3\, aic\right)t + bc\,(i + 2)\right)\lambda\,\delta^2 \\
&+ \left(a^2\,(1 + 2\, i)\, t^2 - ab\,(i + 2)\, t - ac\,(i - 1)\right)\lambda^2\delta + ai\,(at - b)\,\lambda^3
\end{aligned}$$

A similar recurrence can be obtained for $I_{i,2i-1}$. As the previous ones, it is obtained by creative telescoping (see Section 3.3).

3.5 Convolution with Arcs of Circle

We examine the convolution of an arc of circle for power inverse kernels. Though
arcs of circles appear in the literature about convolution surfaces [21, 37], there is
no general formulae for these. In this section we choose a parametrization for arcs
of circle that allows us to write the convolution functions (with varying radius or
scale) in terms of an integral indexed by two integers:

$$F_{i,k}(a, b, c, \lambda, \delta) = \int_{-T}^{T} \frac{(\lambda + \delta t)^k (t^2 + 1)^{i-1}}{(a t^2 - 2 b t + c)^i} dt$$

We then show how to determine closed-form formulae for these integrals, thanks to
some recurrences. We restrict our attention to convolution of arcs of circle with even
power inverse kernels. With odd power inverse kernels, the closed-form formulae for
the convolution function involve elliptic functions and can be rather impractical to
evaluate.

The closed-form formulae for the convolution of arcs of circle with the family
of compact support kernels are challenging to obtain. The software MixedCT[6]
by L. Dumont (in Maple) does meet this challenge. The result would be too
cumbersome to be presented here, and it is not clear at this stage how to use it
efficiently.

3.5.1 Rational Parametrization

When it comes to integration, rational functions are the dependable class [8]. The
main ingredient in obtaining closed-form convolution functions for arcs of circle is
to introduce an appropriate rational parametrization.

We assume that the points O, A, and B are not aligned and such that $|OA| = |OB| = r$. They define a plane in space and two arcs of circle, one of angle α and
the other of angle $\pi + \alpha$ for some $0 < \alpha < \pi$. We have

$$\alpha = \arccos\left(\frac{\overrightarrow{OA} \cdot \overrightarrow{OB}}{r^2}\right) \qquad \text{with} \quad 0 < \alpha < \pi$$

and accordingly to which angle is dealt with we set

$$T = \tan\left(\frac{\alpha}{4}\right) \quad \text{or} \quad T = \tan\left(\frac{\pi + \alpha}{4}\right).$$

---

[6]http://mixedct.gforge.inria.fr

Fig. 3.4 Rational
parametrization of an arc of
circle

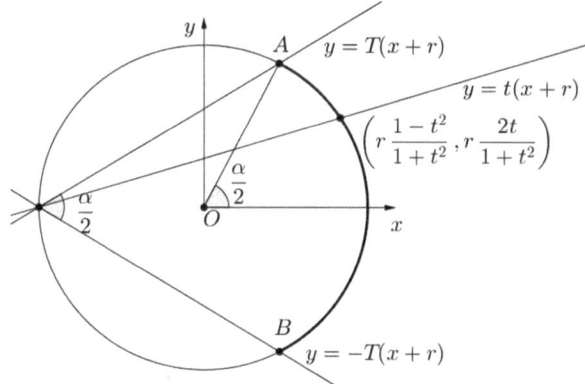

Momentarily we consider the coordinate system (x, y, z) where the origin is the
center of the circle; the x-axis is the bisector of the chosen angle defined by O, A,
and B; and the (x, y) plane is the plane of the circle. See Figure 3.4.

A parametrization of the arc of circle is then given by

$$\Gamma : [-T, T] \longrightarrow \mathbb{R}^3$$

$$t \mapsto \left(r \frac{1 - t^2}{t^2 + 1}, r \frac{2t}{t^2 + 1}, 0 \right).$$

This is obtained by determining the intersection of the circle with the lines of slope
t through the point diametrically opposite to the middle of the arc. Consider a point
$P = (x, y, z)$ in space. We have

$$|P\Gamma(t)|^2 = \frac{\alpha t^2 - 2\beta t + \gamma}{t^2 + 1}$$

where $\quad \alpha = (x + r)^2 + y^2 + z^2, \quad \beta = 2r\, y, \quad \gamma = (x - r)^2 + y^2 + z^2.$

Note that

$$\gamma + \alpha = 2\left(|OP|^2 + r^2\right)$$

$$\alpha\, T^2 + 2\beta\, T + \gamma = (T^2 + 1)\, |AP|^2$$

$$\gamma\, T^2 - 2\beta\, T + \gamma = (T^2 + 1)\, |BP|^2$$

so that (α, β, γ) is actually the solution of a linear system that depends on T and the
squares of the distances of P to O, A, and B. There is a unique solution provided
that A, O, and B are not aligned, i.e., $T(T^2 - 1) \neq 0$. This solution is

$$\alpha = \frac{\left(|PA|^2 + |PB|^2\right)(T^2 + 1) - 4\left(|PO|^2 + r^2\right)}{T^2 - 1},$$

and

$$\beta = \frac{\left(|PA|^2 - |PB|^2\right)\left(T^2 + 1\right)}{T}, \quad \gamma = 2\left(|PO|^2 + r^2\right) - \alpha.$$

3.5.2 Integrals for Convolution

Using the above parametrization of an arc of circle $\overset{\frown}{A^O B}$, the associated convolution function with the power inverse kernel p^{2i} is

$$\mathcal{C}^{2i}_{\overset{\frown}{A^O B}}(P) = \int_{-T}^{T} \frac{1}{|P\Gamma(t)|^{2i}} \frac{2r}{1+t^2} dt = 2r \int_{-T}^{T} \frac{(1+t^2)^{i-1}}{(\alpha t^2 - 2\beta t + \gamma)^i} dt$$
$$= 2r\, \boldsymbol{F}_{i,0}(\alpha, \beta, \gamma, \lambda, \delta)$$

as the infinitesimal arclength is $|\Gamma'(t)| = \frac{2r}{1+t^2}$.

We choose a radius or scale function $\rho, \Lambda : [-T, T] \to \mathbb{R}$ that is linear in the parameter t used above. A more intrinsic choice would be to have a radius or scale function linear in the arclength. Since the arclength is $2r \arctan(t) \sim 2rt + O(t^3)$, linearity in t is a reasonable approximation for arcs defined by an angle less than π. This is illustrated in Figure 3.5.

Convolution with varying radius according to $\rho : t \mapsto \lambda + \delta t$ is then given by

$$\mathcal{H}^{2i}_{\overset{\frown}{A^O B},\rho}(P) = \int_{-T}^{T} \frac{(\lambda + \delta t)^{2i}}{|P\Gamma(t)|^{2i}} \frac{2r}{1+t^2} dt = 2r\, \boldsymbol{F}_{i,2i}(\alpha, \beta, \gamma, \lambda, \delta).$$

Convolution with scale function $\Lambda : t \mapsto \lambda + \delta t$ is given by

$$\mathcal{S}^{2i}_{\overset{\frown}{A^O B},\Lambda}(P) = \int_{-T}^{T} \frac{(\lambda + \delta t)^{2i}}{|P\Gamma(t)|^{2i}} \frac{2r}{1+t^2} \frac{dt}{1+\delta t} = 2r\, \boldsymbol{F}^{-T,T}_{i,2i-1}(\alpha, \beta, \gamma, \lambda, \delta).$$

3.5.3 Closed Forms Through Recurrence Formulae

Given that

$$\boldsymbol{F}_{1,0} = \left[\frac{1}{\sqrt{ca - b^2}} \arctan\left(\frac{at - b}{\sqrt{ca - b^2}} \right) \right]_{-T}^{T}$$

we can recover the expression for $\boldsymbol{F}_{i,0}$, for all $i \in \mathbb{N}$, thanks to the recurrence relationship

$$2\,(i+1)\left(ac-b^2\right)F_{i+2,0} - (1+2\,i)\,(a+c)\,F_{i+1,0}$$

$$+2\,i\,F_{i,0} = \left[\frac{\left(1+t^2\right)^i\left(b\left(t^2-1\right)+(a-c)\,t\right)}{\left(at^2-2\,bt+c\right)^{i+1}}\right]_{-T}^{T}$$

On the other hand, the integrals $F_{i,k}$ satisfy the following recurrence:

$$a\,(k+3)\,F_{i,k+4} - A_3\,F_{i,k+3} + A_2\,F_{i,k+2} + A_1\,F_{i,k+1}$$

$$+A_0\,(k+1)\,F_{i,k} = \left[\frac{\delta^3\left(1+t^2\right)^i(\lambda+\delta\,t)^{k+1}}{\left(at^2-2\,bt+c\right)^{i-1}}\right]_{-T}^{T}$$

where

$$A_3 = -2\,(2\,k+5)\,a\lambda - 2\,b\,(i+2+k)\,\delta,$$

$$A_2 = 6\,(k+2)\,a\lambda^2 + 2\,b\,(5+3\,k+2\,i)\,\delta\,\lambda + ((3+k-2\,i)\,a + (k+2\,i+1)\,c)\,\delta^2,$$

$$A_1 = -2\,a\,(3+2\,k)\,\lambda^3 - 2\,b\,(4+3\,k+i)\,\delta\,\lambda^2$$
$$\quad + 2\,((i-2-k)\,a - (i+1+k)\,c)\,\delta^2\lambda + 2\,b\,(i-2-k)\,\delta^3,$$

$$A_0 = \left(\delta^2+\lambda^2\right)\left(\lambda^2 a + c\delta^2 + 2\,b\lambda\,\delta\right).$$

By specializing the above equation to $k = -1$, one can obtain $F_{i,3}$ from $F_{i,2}$, $F_{i,1}$, and $F_{i,0}$. These latter are thus sufficient to determine $F_{i,k}$ for all $k \geq 4$.

To determine $F_{i,1}$ we observe that, for $i \neq 0$,

$$a\,F_{i+1,1} - F_{i,1} = (b\,\delta + a\,\lambda)\,F_{i+1,0} - \lambda\,F_{i,0} - \frac{\delta}{2\,i}\left[\frac{\left(1+t^2\right)^{\frac{i}{2}}}{\left(at^2-2\,bt+c\right)^{\frac{i}{2}}}\right]_{-T}^{T}$$

and

$$F_{1,1} = \left[\frac{(b\,\delta+a\,\lambda)}{a\sqrt{ac-b^2}}\arctan\left(\frac{at-b}{\sqrt{ac-b^2}}\right) + \frac{\delta}{2a}\ln\left(at^2-2\,bt+c\right)\right]_{-T}^{T}.$$

To determine $F_{i,2}$ we can use the linear recurrence that provides $F_{i,k+2}$ in terms of $F_{i,k+1}$, $F_{i,k}$, $F_{i+1,k}$, and $F_{i+2,k}$. Specialized to $k = 0$ this recurrence simplifies to

$$A_{02}\,F_{i,2} + A_{01}\,F_{i,1} + A_{00}\,F_{i,0} + A_{10}\,F_{i+1,0} + A_{20}\,F_{i+2,0}$$

$$= \left[\delta\,\frac{(\lambda+\delta\,t)\left(1+t^2\right)^i}{\left(at^2-2\,bt+c\right)^{i+1}}\,C\right]_{-T}^{T}$$

where

$$A_{02} = a\left((a-c)\,\delta\,\lambda + b(\delta^2 - \lambda^2)\right),$$

$$A_{01} = -2\,(ib\delta + \lambda\,a)\left(-\lambda^2 b + (a-c)\,\delta\,\lambda + b\delta^2\right),$$

$$A_{00} = -\left(\delta^2 + \lambda^2\right)\left(b\left((2\,i-1)\,a + 2\,ci\right)\delta^2\right.$$
$$\left. + \left((2\,i-1)\,a^2 + ac + 2\,b^2 i\right)\lambda\,\delta + \lambda^2 ab\right),$$

$$A_{10} = (1 + 2\,i)\,a^2\,(a+c)\,\delta\,\lambda^3 + b\left(3\,a^2\,(1+2\,i) + ac\,(3+4\,i) + 2\,ib^2\right)\delta^2\lambda^2$$
$$+ \left((4\,i+1)\,ca^2 + a\left(c^2 + 2\,b^2\,(i+1)\right) + 2\,b^2\,(3\,i+1)\,c\right)\delta^3\lambda$$
$$+ b\left((4\,i+1)\,ca + (1+2\,i)\,c^2 - 2\,ib^2\right)\delta^4,$$

$$A_{20} = -2\,\delta\,(i+1)\,(\lambda\,a + b\delta)\left(ac - b^2\right)\left(a\lambda^2 + c\delta^2 + 2\,b\lambda\,\delta\right),$$

and

$$C = a^2\left(b - bt^2 - (a-c)\,t\right)\lambda^2 + \left(a^2 bt^2 - b^2\,(3\,a+c)\,t + b\left(c^2 + 2\,b^2\right)\right)\delta^2$$
$$+ \left(a^2\,(a-c)\,t^2 - 2\,b\left(b^2 + 2\,a^2 - ac\right)t + b^2\,(3\,a+c)\right)\delta\,\lambda.$$

3.6 Application and Outlook

In this section we illustrate the benefit that arcs of circle provide as basic elements of skeletons. Arcs of circle have desirable approximation power for space curves with high curvature or torsion. The formulae we presented thus allow to use them to generate convolution functions and surfaces: each arc of circle, or line segment, in the skeleton generates a convolution function through the formulae presented in Section 3.5, or 3.4, and the convolution function generated by the skeleton is the sum of all of these. In the examples presented here, we used the kernel p^6 and Hornus formulation (defined in Section 3.2.5.1) to generate the convolution functions. The convolution surfaces presented are the level set at value 0.1 of the obtained convolution functions.

Figure 3.5 shows the convolution with a varying radius of a line segment and four arcs of circles. The skeleton curves all have the same extremities but different radius. The angle supporting the arc of circle thus varies. Only when this angle is close to 2π does one detect visually that the thickness does not vary linearly with the arclength. For each skeleton piece $\Gamma([a, b])$, the radius function $\rho : [a, b] \to \mathbb{R}$ is given by an affine function (in the parameter of the parametrization of Γ) with

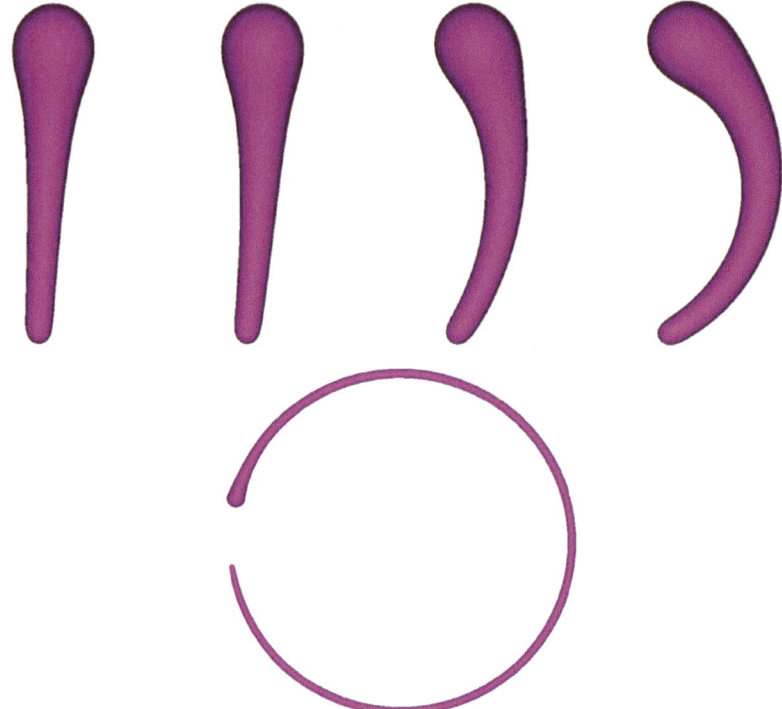

Fig. 3.5 Convolution with varying radius for a line segment (top left) and arcs of circles supported, respectively, by an angle $\frac{\pi}{24}$, $\frac{\pi}{3}$, $\frac{2\pi}{3}$, and $\frac{19\pi}{10}$ (bottom, with a different scale)

$\rho(a) = 1$ and $\rho(b) = 4$. The segments are given by an arclength parametrization of length 25. The arcs are given by the rational parametrization introduced in Section 3.5.1 (Figure 3.4); the radius was chosen to give a chord length of 25 for each example in Figure 3.5.

We can combine several arcs of circles and line segments to model $\mathcal{G}^1$ curves to serve as skeleton. This is illustrated for two closed curves in Figure 3.6.

The widely used approach for more elaborate skeleton curves is to use an approximation by line segments [9, 17, 22, 33, 38]. An issue with this versatile approach is that either the resulting convolution surface presents some visible turns at the joints of line segments or the number of segments must be increased significantly in order to get a visually smooth surface. Convolution for arcs of circles was also examined in [21, 37]. Jin and Tai [21] dealt with planar skeletons. Zanni et al. [37] were geared toward skeleton curves with nonzero torsion: arcs of circles, and line segments, were deformed into helices that have powerful modeling properties for natural shapes, in particular for the animation of hair [3]. The warping technique

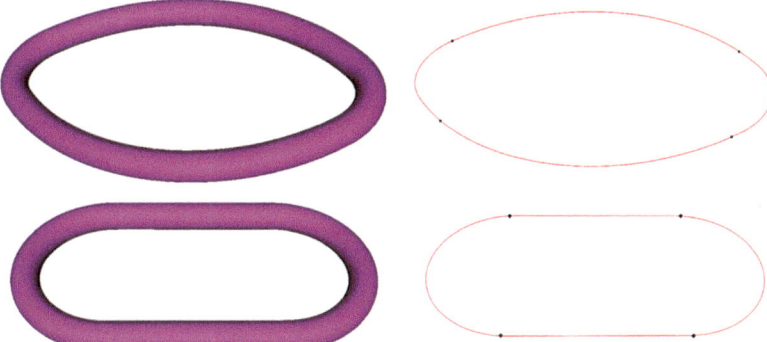

Fig. 3.6 Convolution surface modeling a smooth chain ring. Top row, modeling with arcs of circle only; bottom, arcs of circle and line segments. Left image, the surface; right, the skeleton composed of only four arcs of circle; in black the joint points

Fig. 3.7 Convolution surface around an approximation of the spiral $(\frac{1}{2}t\cos t, \frac{3}{4}t\sin t, \frac{4}{5}t)$, $t \in [0, 2\pi]$. Left image, approximation with 14 arcs of circle; right image, approximation with 14 line segments

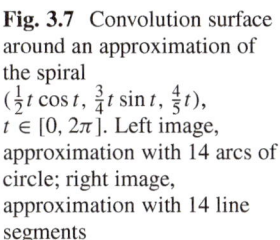
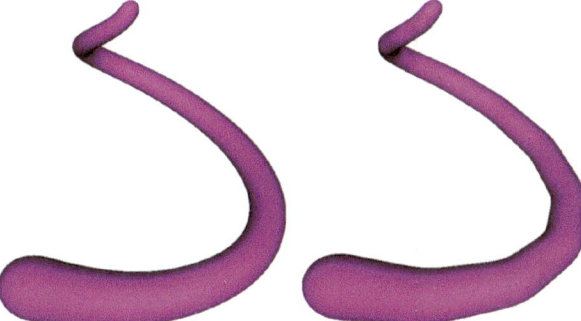

used in [37] allows to decrease substantially the number of skeleton basic elements to be used to obtain a natural-looking shape. This provides a substantial gain on the computational cost as meshing the surface requires the repeated evaluation of the convolution function. Yet the surfaces obtained by warping in [37] exhibit artifacts and singularities so that this technique requires a fine tuning of the warping parameters.

The alternative approach we want to bring forth in this paper is to use a $\mathcal{G}^1$-approximation of the skeleton curve. Arcs of circles have the great advantage to allow the construction of $\mathcal{G}^1$-curves that can approximate any curve [28, 34]. One can thus achieve both mathematically smooth and visually appealing shapes with skeleton curves consisting of few basic elements. This improves visual quality and decreases the computational cost. Figures 3.7, 3.8, and 3.9 compare the convolution surfaces with skeleton curves generated with arcs of circle and line segments. The visual quality of the surface is obtained with much fewer arcs of circles rather than line segments.

Fig. 3.8 Convolution surface around an approximation of the elliptical helix $(2\cos t, 3\sin t, t)$, $t \in [0, 6\pi]$. Top image, approximation with 42 arcs of circle; bottom image, approximation with 42 line segments

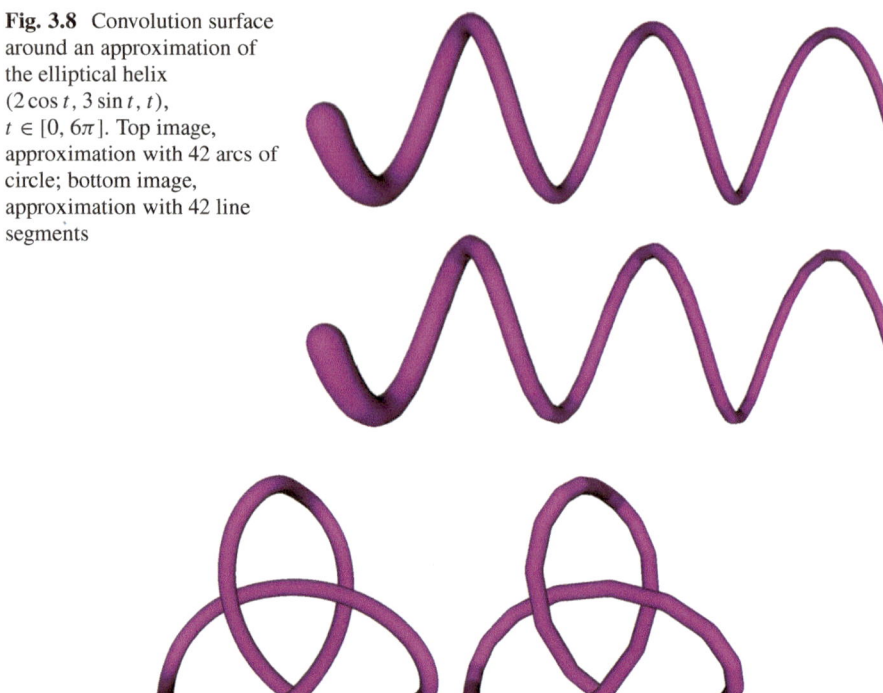

Fig. 3.9 Convolution surface around an approximation of the closed curve $(-10\cos t - 2\cos 5t + 15\sin 2t, -15\cos 2t + 10\sin t - 2\sin 5t, 10\cos 3t)$, $t \in [0, 2\pi]$. Left image, approximation with 34 arcs of circle; right image, approximation with 34 line segments

3.7 Conclusion

In this paper we have focused on convolution surfaces along skeleton consisting of a single curve, either open or closed. We propose to use approximation by line segments and arcs of circles to approximate this curve so as to obtain quality convolution surfaces with lower computational cost. To this effect we provided explicit formulae for the convolution functions for both line segments and arcs of circles, with varying radii or scale functions. These formulae have great generality that draws from the use of recurrence formulae that were obtained with a new technique, creative telescoping.

The great advantage of convolution is to provide a practical mathematical definition of a smooth surface around complex skeletons made of intersecting curves and surfaces. Contrary to offset, sweep, or canal surfaces, convolution surfaces naturally blend smoothly multiple primitive shapes. There are nonetheless challenges in their use. First the visualization mostly relies on refined marching cube algorithms. An alternative approach would be based on the prior construction of a *scaffold* around the skeleton as introduced in [30], with an alternative approach in [15]. A second challenge is the control of the topology and geometry. This problem was tackled in [26, 36]. We expect to provide an alternative more intrinsic formulation, with mathematical guarantees.

Acknowledgements This project has received funding from the European Union's Horizon 2020 research and innovation programme under the Marie Skłodowska-Curie grant agreement No 675789. We want to thank Shaoshi Chen, Bruno Salvy, Alin Bostan, Louis Dumont, Frédéric Chyzak, and Christoph Koutschan for several enlightening discussions on creative telescoping over the years, as well as making their fantastic software available.

References

1. Alexe, A., Barthe, L., Cani, M.P., Gaildrat, V.: Shape modeling by sketching using convolution surfaces. In: 13th Pacific Graphics Short papers, Macau, China (2005)
2. Bernhardt, A., Pihuit, A., Cani, M.P., Barthe, L.: Matisse : painting 2D regions for modeling free-form shapes. In: Eurographics Workshop on Sketch-Based Interfaces and Modeling (SBIM), pp. 57–64. Annecy, France (2008)
3. Bertails, F., Audoly, B., Cani, M.P., Querleux, B., Leroy, F., Lévêque, J.L.: Super-helices for predicting the dynamics of natural hair. ACM Trans. Graph. **25**(3), 1180 (2006)
4. Blinn, J.F.: A generalization of algebraic surface drawing. ACM Trans. Graph. **1**(3), 235–256 (1982)
5. Bloomenthal, J., Shoemake, K.: Convolution surfaces. ACM SIGGRAPH Comput. Graph. **25**(4), 251–256 (1991)
6. Bostan, A., Chen, S., Chyzak, F., Li, Z., Xin, G.: Hermite reduction and creative telescoping for hyperexponential functions. In: ISSAC 2013—Proceedings of the 38th International Symposium on Symbolic and Algebraic Computation, pp. 77–84. ACM, New York (2013)
7. Bostan, A., Dumont, L., Salvy, B.: Efficient algorithms for mixed creative telscoping. In: Proceedings of the ACM on International Symposium on Symbolic and Algebraic Computation, ISSAC '16, pp. 127–134. ACM, New York, NY (2016)
8. Bronstein, M.: Symbolic Integration I. Algorithms and Computation in Mathematics, vol. 1. Springer, Berlin (1997)
9. Cani, M.P., Hornus, S.: Subdivision-curve primitives: a new solution for interactive implicit modeling. In: Proceedings - International Conference on Shape Modeling and Applications, SMI 2001, pp. 82–88 (2001)
10. Cani, M.P., Igarashi, T., Wyvill, G.: Interactive Shape Design. Synthesis Lectures on Computer Graphics and Animation. Morgan & Claypool Publishers, San Rafael, CA (2008)
11. Chen, S., Kauers, M.: Trading order for degree in creative telescoping. J. Symb. Comput. **47**(8), 968–995 (2012)
12. Chen, S., Kauers, M.: Some open problems related to creative telescoping. J. Syst. Sci. Complex. **30**(1), 154–172 (2017)
13. Chyzak, F.: An extension of Zeilberger's fast algorithm to general holonomic functions. Discret. Math. **217**(1), 115–134 (2000)
14. Chyzak, F., Salvy, B.: Non-commutative elimination in ore algebras proves multivariate identities. J. Symb. Comput. **26**(2), 187–227 (1998)
15. Fuentes Suárez, A.J., Hubert, E.: Scaffolding skeletons using spherical Voronoi diagrams. Electron Notes Discrete Math. **62**, 45–50 (2017)
16. He, Q., Tong, M., Liu, Y.: Face modeling and wrinkle simulation using convolution surface. In: Perales, F.J., Fisher, R.B. (eds.) Articulated Motion and Deformable Objects. AMDO 2006. Lecture Notes in Computer Science, vol. 4069, pp. 244–251. Springer, Berlin (2006)

17. Hornus, S., Angelidis, A., Cani, M.P.: Implicit modeling using subdivision curves. Vis. Comput. **19**(2), 94–104 (2003)
18. Hubert, E.: Convolution surfaces based on polygons for infinite and compact support kernels. Graph. Model. **74**(1), 1–13 (2012)
19. Hubert, E., Cani, M.P.: Convolution surfaces based on polygonal curve skeletons. J. Symb. Comput. **47**(6), 680–699 (2012)
20. Jin, X., Tai, C.L.: Analytical methods for polynomial weighted convolution surfaces with various kernels. Comput. Graph. **26**, 437–447 (2002)
21. Jin, X., Tai, C.L.: Convolution surfaces for arcs and quadratic curves with a varying kernel. Vis. Comput. **18**(8), 530–546 (2002)
22. Jin, X., Tai, C.L., Feng, J., Peng, Q.: Convolution surfaces for line skeletons with polynomial weight distributions. ACM J. Graph. Tools **6**(3), 17–28 (2001)
23. Jin, X., Tai, C.L., Zhang, H.: Implicit modeling from polygon soup using convolution. Vis. Comput. **25**(3), 279–288 (2009)
24. Koutschan, C.: A fast approach to creative telescoping. Math. Comput. Sci. **4**, 259–266 (2010)
25. Koutschan, C.: Creative telescoping for holonomic functions. In: Schneider, C., Blümlein, J. (eds.) Computer Algebra in Quantum Field Theory. Texts & Monographs in Symbolic Computation (A Series of the Research Institute for Symbolic Computation, Johannes Kepler University, Linz, Austria), pp. 171–194. Springer, Vienna (2013)
26. Ma, G., Crawford, R.H.: Topological consistency in skeletal modeling with convolution surfaces. In: Volume 3: 28th Computers and Information in Engineering Conference, Parts A and B, pp. 307–315. ASME, New York (2008)
27. Milnor, J.W.: Morse Theory, p. 153. Princeton University Press, Princeton (1963)
28. Nutbourne, A.W., Martin, R.R.: Differential Geometry Applied to Curve and Surface Design. John Wiley & Sons, Hoboken, NJ (1988)
29. Ore, O.: Theory of non-commutative polynomials. Ann. Math. **34**(3), 480 (1933)
30. Panotopoulou, A., Welker, K., Ross, E., Hubert, E., Morin, G.: Scaffolding a Skeleton (2018)
31. Piessens, R., de Doncker-Kapenga, E., Überhuber, C.W., Kahaner, D.K.: Quadpack. Springer Series in Computational Mathematics, vol. 1. Springer, Berlin (1983)
32. Sherstyuk, A.: Interactive shape design with convolution surfaces. In: Shape Modeling International '99, pp. 56–65 (1999)
33. Sherstyuk, A.: Kernel functions in convolution surfaces: a comparative analysis. Vis. Comput. **15**(4), 171–182 (1999)
34. Song, X., Aigner, M., Chen, F., Jüttler, B.: Circular spline fitting using an evolution process. J. Comput. Appl. Math. **231**(1), 423–433 (2009)
35. Wyvill, G., McPheeters, C., Wyvill, B.: Data structure for soft objects. Vis. Comput. **2**(4), 227–234 (1986)
36. Zanni, C.: Skeleton-based implicit modeling & applications. Phd, Université de Grenoble (2013)
37. Zanni, C., Hubert, E., Cani, M.P.: Warp-based helical implicit primitives. Comput. Graph. **35**(3), 517–523 (2011)
38. Zanni, C., Bernhardt, A., Quiblier, M., Cani, M.P.: SCALe-invariant integral surfaces. Comput. Graph. Forum **32**(8), 219–232 (2013)
39. Zhu, X., Jin, X., Liu, S., Zhao, H.: Analytical solutions for sketch-based convolution surface modeling on the gpu. Vis. Comput. **28**(11), 1115–1125 (2012)
40. Zhu, X., Jin, X., You, L.: High-quality tree structures modelling using local convolution surface approximation. Vis. Comput. **31**(1), 69–82 (2015)
41. Zhu, X., Song, L., You, L., Zhu, M., Wang, X., Jin, X.: Brush2model: convolution surface-based brushes for 3D modelling in head-mounted display-based virtual environments. Comput. Anim. Virtual Worlds **28**(3–4), e1764 (2017)

Chapter 4
Exploring 2D Shape Complexity

Erin Chambers, Tegan Emerson, Cindy Grimm, and Kathryn Leonard

Abstract In this paper, we explore different notions of shape complexity, drawing from established work in mathematics, computer science, and computer vision. Our measures divide naturally into three main categories: skeleton-based, symmetry-based, and those based on boundary sampling. We apply these to an established library of shapes, using k-medoids clustering to understand what aspects of shape complexity are captured by each notion. Our contributions include a new measure of complexity based on the Blum medial axis and the notion of *persistent complexity* as captured by histograms at multiple scales rather than a single numerical value.

4.1 Introduction

Quantifying shape complexity and similarity has a rich history in many fields of mathematics and computer science, including work in fields such as differential geometry, topology, computational geometry, computer graphics, and computer vision. This paper draws measures common to several of those fields in order to compare and contrast their ability to capture complexity of a variety of two-dimensional shapes, as well as introducing a new approach to complexity based on measures taken along the Blum medial axis.

E. Chambers (✉)
Department of Computer Science, St Louis University, St Louis, MO, USA
e-mail: erin.chambers@slu.edu

T. Emerson
Colorado State University, Fort Collins, CO, USA
e-mail: emerson@math.colostate.edu

C. Grimm
Oregon State University, Corvallis, OR, USA
e-mail: cindy.grimm@oregonstate.edu

K. Leonard
Occidental College, Department of Computer Science, Los Angeles, CA, USA
e-mail: kleonard.ci@gmail.com

© The Author(s) and the Association for Women in Mathematics 2018
A. Genctav et al. (eds.), *Research in Shape Analysis*, Association for Women in Mathematics Series 12, https://doi.org/10.1007/978-3-319-77066-6_4

4.1.1 Prior Work on Shape Complexity

Several examples of prior work on shape complexity draw from the information theoretic framework. For example, in [4], the authors consider a notion of complexity based on computing measures related to the Kolmogorov complexity of a set of sample points from the boundary of the contour. This method has the advantage of calculating a single complexity value which in practice does well at distinguishing between various classes of shapes, in agreement with a user study. Unfortunately, it is extremely difficult to classify exactly what is being measured given the loose connection to Kolmogorov complexity, which is uncomputable.

Similar work in computer vision has taken a different approach, quantifying complexity by how difficult it is to capture or cover the shape by simpler ones. For example, in [17], the authors measure complexity by attempting to determine how many ellipses are necessary to cover a 2D shape. They compute the medial axis for each edge, prune to reduce noise, and then calculate an entropy in order to determine how many ellipses are necessary to cover the total area as closely as possible. The basic premise is that a shape requiring more ellipses, or one resulting in less coverage with more ellipses, will correspond to a more complex shape.

In [3], the authors focus on a classical computational geometric measure of complexity, namely, deciding how quickly they can triangulate a given input polygon. While of theoretical interest, this measure does not translate to a richer notion of shape complexity.

Curvature of the boundary appears in several complexity measures. In [5], the authors determine which regions of a particular contour are more "surprising" in a probabilistic sense, finding that regions of negative curvature carry more information than those of positive curvature of the same magnitude. Similarly, [16] defines probability distributions of curvatures in order to assign an entropy to each shape. A discretization-independent approach of the same idea can be found in [18]. In [9], total curvature of the boundary gives the adaptive codelength for a contour, where one may view codelength as a proxy for complexity. One drawback to the curvature-based derivations, however, is their lack of robustness to noise or small-scale detail on the boundary which can dramatically alter curvature distributions.

Work that attempts to classify the complexity of three-dimensional objects is also worth mentioning, even though our work focuses on the 2D setting. In [15], for example, the authors propose a signature for an object as a shape distribution, which is a similar idea to how we approach the problem in the 2D setting, but their shape function primarily measures local geometric properties of an object. Similarly, in [20] the authors use curvature-based and information theoretic measures to classify 3D shapes, although in this work they focus on shape segmentation and identification of prominent features, while our work focuses on classifying overall shape complexity.

4.1.2   Our Contribution

The purpose of this work is to lay the foundation for a definition, or collection of definitions, of shape complexity that captures the full range of its natural dimensions. To do so, we explore several classical complexity measures and introduce a skeleton-based measure. Our main contribution is the idea of a persistent measure of shape complexity, which examines complexity as a shape is eroded into its least complex approximation. Complexity is then captured by a collection of values or histogram rather than a single value.

The measures we implement below divide naturally into three categories: skeleton-based, symmetry-based, and boundary sampling measures. For the boundary approach, we extract measures as the boundary is downsampled, which allows us to differentiate persistent from transient sources of complexity.

We begin with Section 4.2, which establishes the context for our exploration and provides an initial foray into general principles of shape complexity. Section 4.3 describes the complexity measures we use, with results from clustering based on those measures displayed in Section 4.4. Finally, we discuss our results and future directions to explore in Section 4.5.

4.2   Defining Complexity

As with many attributes assigned to shapes, the complexity of a given shape seems intuitively straightforward but computationally elusive. We believe a few principles should be satisfied in any reasonable definition of 2D shape complexity:

1. A circle should have the minimum complexity.
2. Adding a part to a shape that is different from all existing parts should increase complexity.
3. A shape with parts that are self-similar should be less complex than a shape with the same number of parts where the parts are dissimilar.
4. Shapes, together with a complexity measure, should be a partially ordered set.

The above form an objective measure for shape complexity, independent of a particular setting. In addition to exploring objective measures, we are also interested in definitions of complexity that are useful in a particular setting. The next section explores the idea of domain-dependent complexity and how that might result in different definitions or representations.

4.2.1 Qualitative Complexity

We outline here two possible *qualitative* methods of defining what complexity means and how those methods might relate to specific tasks or applications. In this we take the view that defining a complexity measure is not completely independent of what you are using that measure for. For example, one measure might capture how difficult a shape is to build, while another might better capture how difficult it is for a human to recognize or classify a set of shapes.

4.2.1.1 Construction Definition

This definition derives from the idea of procedural modeling [14, 19]—specifying a sequence of operations that create the shape in question. Complexity is a measure of how deep the tree of operations has to be; conversely, the further down the tree you go, the more complex the object. Obviously, this depends on the set of available operators. We outline here a plausibly complete set of operators and a few examples of how a shape could be constructed from those operators. Note that there may not be a single, unique set of operations for each shape.

Potential tasks or applications include shape representation, modeling, functional shape comparison, and part decomposition. This is also a *representational* definition, in that the encoding of the shape as a tree of operations means that the shape can be recovered from the complexity measure. The definition is also partially *semantic*, in that the operators form a language for describing shape.

The operators are illustrated in Figure 4.1. The assumption is that the base shape is a circle. The operators are:

- Protrusions: Add one (or a pattern of) protrusions/extrusions.
- Pockets: Add one (or a pattern of) pockets or indents

Fig. 4.1 A possible set of construction operators. This set is complete but possibly overcomplete

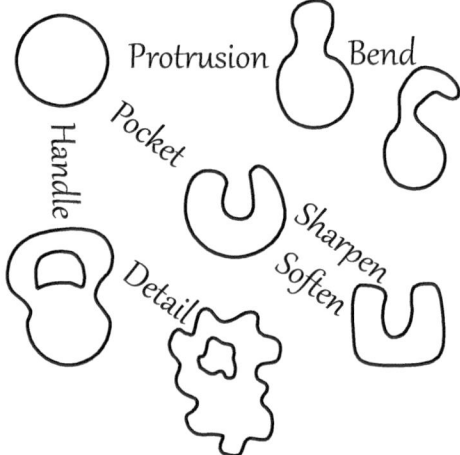

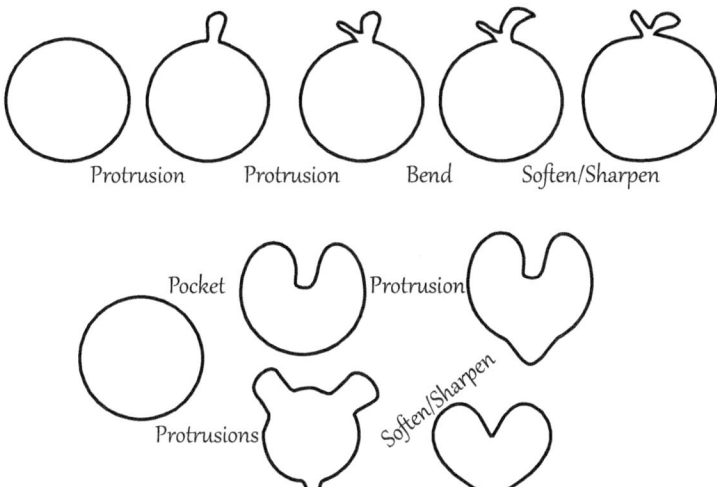

Fig. 4.2 Example construction sequence of two shapes from our data set. Note that the construction order need not be unique—the heart top, for example, can be seen as either a pocket or two protrusions

- Add handle: Add a handle/join a protrusion to the shape/close off a pocket.
- Bends: Bend or curve a protrusion or pocket to reshape it.
- Sharpen or soften: Change the curvature along the boundary to produce a feature (such as a corner or edge).
- Detail pattern: Add a regular or semi-regular pattern to some (or all) of the boundary at a small scale (relative to the overall size of the shape).

In Figure 4.2 we show an example of applying the operators to produce one of our test shapes. Note that there is a natural mapping between the operators and skeleton editing.

4.2.1.2   Predictability Definition

This definition is very perceptual or experience based. The essential notion is that complex shapes both surprise us and require more thought or effort to remember well enough to reproduce. For example, although a drawing of a horse is fairly complex from a total curvature or part standpoint, it is easily mapped to a canonical image of a horse in our head. Similarly, a shape might consist of self-similar or repeating patterns (e.g., Julia set or a symmetric shape with a pattern added to the boundary); in which case, while the geometry is complex, the mental image of it is not. On the other hand, a random pattern (such as the blob in the top row of Figure 4.11) might be fairly simple geometrically but complex because it does not map to any template image in our head.

Essentially, the complexity is a measure of how much the given shape deviates from a canonical set of template shapes. Possible applications or tasks related to this definition are template matching, shape classification, and stability of shape.

4.2.2 Evaluating Quantitative Measures of Complexity

In the previous sections, we presented both existing quantitative, global measures of complexity from the literature and two hypothetical qualitative measures of complexity. In Section 4.3 we will define several families of complexity measures using the notion of sampling and resulting histograms. The questions that link our qualitative definitions to our quantitative ones are: How do we identify quantitative measures that match our qualitative measures of complexity? How do measures differ in how they define complexity? In this work, we use clustering as a first step to answering these questions.

It is likely that complexity is not a 1D, well-defined metric, which means that simply ranking shapes from low complexity to high complexity with one global measure may not be appropriate, particularly if we want to use multiple measures to capture different aspects of complexity. We therefore perform unsupervised clustering on a particular measure or set of measures. An effective measure, or set of measures, is one which consistently groups shapes with certain recognizable complexity features with a recognizable similarity in complexity level.

In the following section, we describe our shape data set and define our complexity measures. In Section 4.4 we describe our clustering algorithm and display the resulting clusters.

4.3 Measures

Given a particular quantitative measure, we can apply it either to the entire shape to obtain a global measure or to the shape at different sampling rates to determine the scales at which certain values of the measure persist. We take as global measures four quantities derived from the medial axis representation [9, 11, 13], which are computed on each medial branch, and three quantities measured on the boundary curve. The measures taken at different sampling rates, so-called sampling measures, are quantities that capture the *changes* in the shape as the resolution of the shape is decreased. We employ two methods for downsampling the shape: boundary point subsampling and image-based subsampling.

Although the boundary of the shape and its interior are linked, some measures focus primarily on measuring how the boundary changes (e.g., curvature), while others are more concerned with the interior shape (covering the shape with a set of ellipses). Some, such as the medial axis measures, capture both. We store the sampling measures and the medial axis measures taken across branches as histograms ($n = 10$ bins), rather than computing statistics such as the mean or measuring entropy [4], to preserve the variation in the data.

4.3.1 Shape Database

We use the MPEG-7 database [1] taking one shape from each of the 70 classes, as well as all shapes in a device class that allows us to understand how complexity varies as a very regular shape is perturbed in different ways. All shapes are represented as a single closed curve, ignoring interior detail. The original images are 256 by 256 with the boundary curves extracted for [10]. This results in around 500 points per shape. We resample the boundary curves using arclength parameterization and normalize the scale of the shape to fit in the unit square.

To calculate the curvatures, we use the formulation in [12] which is based on the angle change over the length. The arclength parameterization and scaling to the unit square together keep the curvatures in roughly the range -100 to 100.

4.3.2 Global Measures

We include four global measures from the literature: ratio of perimeter length to area, total curvature [9], object symmetries [13], and salience measures on the medial axis such as erosion thickness and shape tubularity [11]. From the domain-independent standpoint, the ratio of perimeter length to area, symmetries, and the medial axis measures capture in various ways how far a shape has deviated from a circle. The medial axis and symmetry measures also capture self-similarity of parts. From the domain-dependent standpoint, the total curvature captures the bending and sharpen/soften operations; the ratio of perimeter length to area captures protrusions, pockets, and handles; and the medial axis measures capture an aspect of each of the operations. Predictability is captured in a simplistic way by object symmetries but requires a full probabilistic study of shape occurrence that we do not take up here.

4.3.2.1 Skeleton Measures

We compute the Blum medial axis corresponding to the boundaries of the shapes at the full sampling rate and compute several salience measures on the axis. The extended distance function (EDF) measures the depth of each medial point within the shape, where depth is defined as the second longest geodesic distance along the axis from a given point to the boundary curve [11]. The weighted extended distance function (WEDF) measures another kind of depth, where the depth is the area of the part of the shape subtended by a particular medial point [8]. The erosion thickness (ET) measures how blob-like a shape is at a particular point [11], and the shape tubularity (ST) measures how tubelike it is [11]. The values of these measures vary continuously across primary branch points but will exhibit discontinuities moving from a primary branch to a secondary branch. In this way, the measures taken together provide a picture of the morphology and relative size of the various shape parts comprising the shape [10]. See Figure 4.3 for an example of a hand contour and the variation of its ST, ET, WEDF, and EDF values.

Fig. 4.3 Heat maps showing
values of measures taken
from the medial axis. From
left to right: EDF, a
length-based depth; WEDF,
an area-based depth; ET, a
measure of how blob-like a
region is; and ST, a measure
of how tubelike a region is

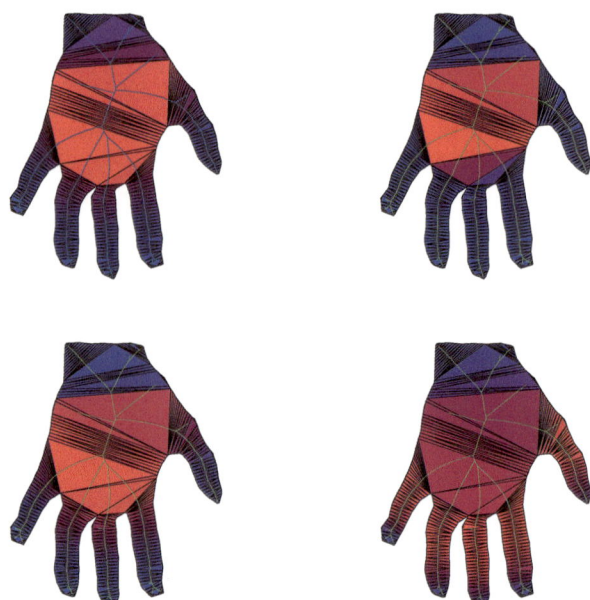

We sample the WEDF, EDF, ST, and ET at medial branch points and their
neighbors, as these are the locations of the discontinuities that characterize the
shape parts. We normalize the histograms of ET and ST but leave the WEDF and
EDF histograms as raw counts in order to preserve information about the number
and types of branches within the shape. Note that when we combine all measures
together, we normalize EDF and WEDF to prevent their values from dominating the
clustering.

4.3.3 Sampling-Based Measures

We employ two sampling approaches, one based on downsampling the boundary
itself and the other downsampling the area enclosed by the boundary. We first
describe the downsampling approaches and then the measures we calculate on the
downsampled data. In general, boundary downsampling reduces the complexity of
the boundary by reducing the number of points used to represent it (**not** smoothing),
while area downsampling uses bigger and bigger blocks to represent the area
coverage of the shape.

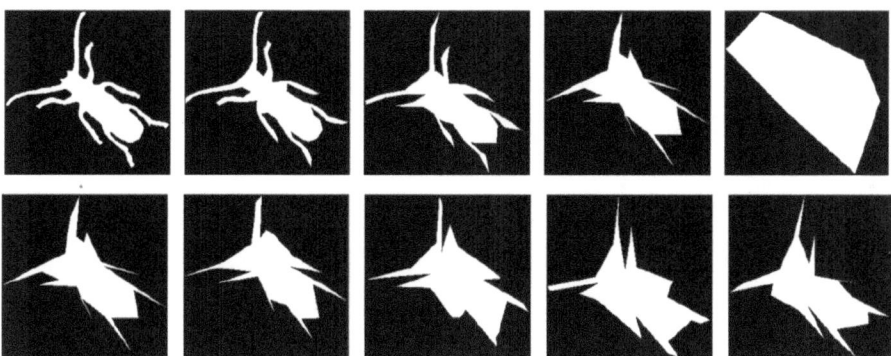

Fig. 4.4 Downsampling the boundary. Top row, from left to right: decreasing the number of points used to represent the boundary (500, 100, 50, 25, and 8). Bottom row: shifting the starting point by 4 for the fourth image above (5 shifts total)

Table 4.1 Sampling rates. The grid pixel count does not include an extra padding of 1 pixel around the boundary.

Boundary					
No. points	500	100	50	25	8
No. shifts	1	2	3	5	16
Grid					
No. pixels	128 × 128	64 × 64	32 × 32	16 × 16	
No. shifts	1	2	4	8	

4.3.3.1 Boundary Downsampling

To downsample the boundary, we linearly approximate the boundary using a reduced number of points (5 levels: 500, 100, 50, 25, and 8). For our shapes this represents the spectrum from roughly full resolution to a convex shape (see Figure 4.4). We use arclength sampling, dividing the original curve into equal-length segments. Because the starting point influences the downsampled shape, we compute multiple downsampled versions, each with a shifted starting point (by 4 points). We then average (values) or sum (histograms) the measures as appropriate for each of the downsampled versions at that level. Table 4.1 summarizes the downsampling values.

When downsampling we also record which segment corresponds to each original point in order to calculate the distance from the original points to the downsampled curve.

4.3.3.2 Area Downsampling

We begin by scan-converting the original boundary curve into a 256×256 image I with 16 pixels of padding on all sides. We downsample by placing a grid with an n pixel neighborhood ($n \in [2, 4, 8, 16]$) on top of I. We store two values per grid

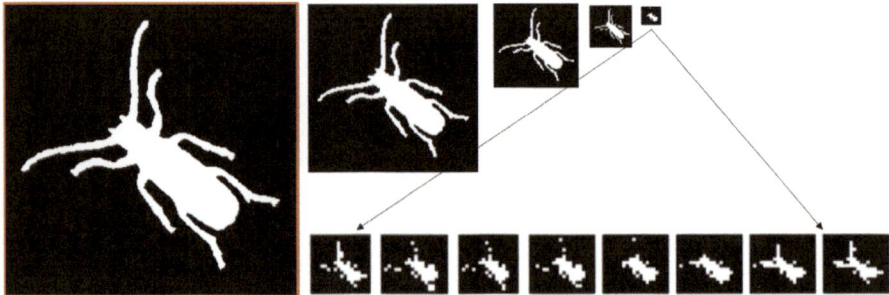

Fig. 4.5 Downsampling using a grid. The largest image is the original 256×256 image. Images are downsampled, with a shift added (all eight images for the lowest resolution are shown enlarged)

cell: a binary value indicating whether or not the grid cell overlaps I at all and the percentage of the overlap. As in the boundary case, we shift the grid starting point (2 pixels in each direction) and combine the results (see Table 4.1). Resulting images are shown in Figure 4.5.

4.3.3.3 Boundary Measures

We normalize the measures, where appropriate, by dividing by the same quantity associated to the original, fully sampled shape. Continuous values are stored as histograms. Denote by d_i the distance from each of the original points to the closest point on the corresponding segment of a downsampled boundary. Three measures use the Voronoi diagram and Delaunay triangulation, which are computed from the downsampled boundary. More precisely, we calculate:

- Length of downsampled boundary/Length of original boundary. This captures the depth of protrusions and pockets and how many there are.
- Area enclosed by the downsampled boundary/area of original boundary. This captures the area of protrusions and pockets.
- L^2 norm on the approximation error produced by downsampling: mean of d_i. This captures the average distance of the original shape boundary from successively simpler (more convex) versions of itself.
- Hausdorff norm on the approximation error produced by downsampling: max of d_i. This captures the maximum deviation of the original shape boundary from successively simpler (more convex) versions of itself.
- Distribution of d_i (histogram). Bins are evenly spaced between 0 and 0.1 and then at 0.5 and the maximum of d_i. This captures the variability in distances between the original shape and successively simpler versions of itself.
- Distribution of curvature (histogram). Bins are evenly spaced *after* taking arctan (so-called Shape Index [7]). This captures the persistence of high curvature regions as the shape is simplified.

- Distribution of edge lengths in the Voronoi diagram. This provides a measure of the average aspect ratios of shape parts.
- Distribution of triangle areas in the Delaunay triangulation. This captures the percentage of smaller components to larger components in the shape.
- Percentage of Voronoi cell centers that lie inside the shape versus outside. This is a measure of how non-convex a shape remains as it is simplified.

The boundary measures, in general, provide information about the size, number, shape, and depth of protrusions. By using downsampling we can additionally capture how "stable" those protrusions are—i.e., how quickly they disappear.

4.3.3.4 Moment Measures

In order to capture a coarse self-similarity measure, we compute the seven complex moment invariants described in [13] at the different boundary downsampling scales. Taken together, these moments capture all rotational symmetries. Note that these will not reflect the addition operations such as protrusions, as long as they are added along all axes of symmetry; however, any additions that are not symmetric will change the moment measures quite drastically. In other words, the moment measures capture an aspect of the self-similarity of the parts comprising a shape, one of our domain-independent complexity tests.

4.3.3.5 Area Measures

We store the area of the occupied downsampled image over the occupied area in the original image. Note that this measure tends to increase, while the area based on downsampling the boundary tends to decrease and is potentially better at capturing the shape of pockets versus protrusions and also their alignment with respect to the overall shape. We also store the distribution of occupied pixels as a histogram.

4.4 Clustering

In order to interpret what the different complexity measures capture, we perform k-medoids clustering using each set of measures as feature vectors. K-medoids is a distance-based, unsupervised learning algorithm that groups the data set into k clusters. Unlike its sister, the k-means algorithm, k-medoids requires that the centroid of each cluster (the point to which distances are being computed and compared) be a sample from the data set. In many applications this forcing of the centroid to be a sample from the data allows for more significant interpretations.

The standard Euclidean distance is used to determine cluster membership in the clustering results. For all clusterings illustrated, we have let $k = 6$. This number was determined based on comparing the average within-cluster distance sum across clusterings produced for k ranging from 1 to 15. When k is equal to the number of points in the data set, the within-cluster distance sum is zero since each point is its own centroid. Reasonable values of k can be identified by looking for an elbow in the graph of average within-cluster distance sum versus k. For the different feature vectors considered, the elbow occurred between $k = 3$ and $k = 7$ in the majority of trials. For the sake of comparing the results across different feature vectors more smoothly, we fixed $k = 6$. We note that there are more complex ways of identifying "optimal" cluster numbers that consider both the within-cluster distance and across-cluster separation. However, for our purposes the within-cluster measure appears sufficient. In future work a more thorough analysis may be performed, allowing for different numbers of clusters for each set of feature vectors.

Results of the clusterings shown herein are based on running 50 trials of k-medoids for each feature vector set and selecting the one with the smallest within-cluster distance sum. Figures 4.8, 4.12, 4.14, 4.10, 4.6, 4.16, and 4.18 show the six different medoids chosen for each of the feature vector sets. These medoids are color-coded to provide cluster membership information in the full clustering results shown in Figures 4.9, 4.13, 4.15, 4.11, 4.7, 4.17, and 4.19. In these figures the medoid of each cluster is the first shape for each new cluster color and has been given a black border. The subsequent shapes in a given color are ordered from closest to farthest from the medoid as measured by Euclidean distance.

The different sets of feature vectors considered are produced as follows. We group the features according to their appearance in Section 4.3.

- Global measures

  - **Non-skeleton global feature vectors:** The global clusters use a feature vector that is 12-dimensional. Included in the feature vector are the seven moment measures computed at the full scale of the image and also boundary length-to-area ratio, total curvature, curvature entropy, angular entropy, and distance to original entropy. This set of feature vectors is based exclusively on the complete sampling of the shape and complexity values that can be found in existing literature. Figures 4.6 and 4.7 show the cluster centroids and entire clustering based on the global feature vectors.
  - **Skeleton feature vectors:** The skeleton feature vectors are composed of histograms of the ET and ST values at all branch points and neighbors of branch points. Separate histograms are produced for ET, ST, and each point type. The ET/ST values are all normalized relative to the max ET/ST for the specific shape, and then the histogram is normalized based on the total number of points (branch or neighbor points) being considered. EDF and WEDF are also computed for branch points and neighbor points. The values are scaled relative to the maximum value of EDF/WEDF, but the number of points is not normalized in order to preserve information about the number of branches in each bin. This results in a 40-dimensional skeleton-based feature vector. The cluster centroids for the clustering built on the skeleton feature vectors are shown in Figure 4.8, and the full clustering is shown in Figure 4.9.

Fig. 4.6 Medoids from clustering based on global measures

Fig. 4.7 Clustering based on global measures

Fig. 4.8 Medoids from clustering based on skeleton measures

Fig. 4.9 Clustering based on skeleton measures

- Sampling measures

  - **Boundary feature vectors:** Boundary features are all computed at the five
    different downsampling scales. Four values are histogrammed into ten bins for
    each value as described in Section 4.3.3.3. Another four values are computed
    at each scale, also listed in Section 4.3.3.3. Consequently, the boundary feature
    vectors are 220-dimensional (5 scales × (4 histogrammed values × 10 bins
    per histogram + 4 non-histogrammed values) = 220). Results of clustering
    on the boundary feature vectors and the cluster centroids are contained in
    Figures 4.11 and 4.10, respectively.

  - **Moment feature vectors:** The moment measures considered are the seven
    values described in [13]. These seven values are computed at the five different
    boundary downsampling scales. This results in a 35-dimensional feature
    vector for each shape that we use to cluster on. Cluster centroids for clustering
    on the moment measure feature vectors are shown in Figure 4.12, and the
    complete clustering is shown in Figure 4.13.

  - **Coverage feature vectors:** The coverage feature vector is 44-dimensional and
    consists of one non-histogrammed value (area change) and one histogrammed
    value (pixel occupied distribution, 10 bins) for each of the four downsam-
    plings. Centroids for each cluster and the full clustering based on the coverage
    measure feature vectors are provided in Figures 4.14 and 4.15, respectively.

- Concatenation of global and sampling measures

  - **With moments:** Clusterings based on concatenated sets of feature vectors are
    shown in Figures 4.16, 4.17, 4.18, and 4.19. Figures 4.16 and 4.17 show the
    medoids and clustering, respectively, where boundary, coverage, moment, and
    normalized skeleton feature vectors are concatenated.

  - **Without moments:** Alternatively, the results based on clustering boundary,
    coverage, and skeleton feature vectors (leaving out moment feature vectors)
    produce the medoids and clusters in Figures 4.18 and 4.19. The concatenated
    feature vectors built by combining the boundary, moment, coverage, and
    skeleton feature vectors are 339-dimensional, while those without the moment
    feature vector are 304-dimensional.

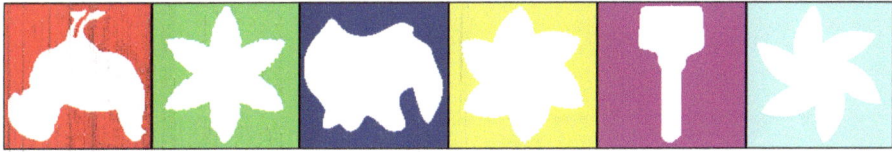

Fig. 4.10 Medoids from clustering based on boundary measures

Fig. 4.11 Clustering based on boundary measures

Fig. 4.12 Medoids from clustering based on moments measures

Fig. 4.13 Clustering based on moments measures

Fig. 4.14 Medoids from clustering based on area measures

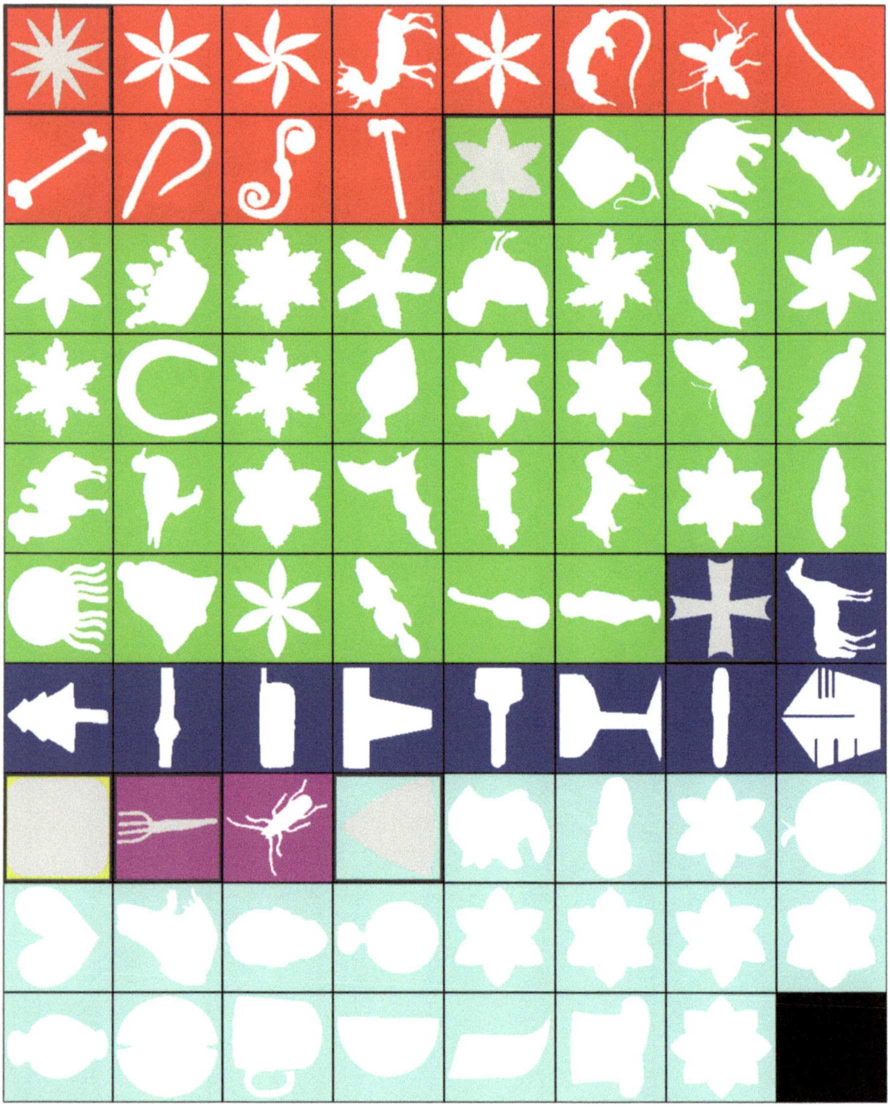

Fig. 4.15 Clustering based on area measures

Fig. 4.16 Medoids from clustering based on concatenated boundary, moment, coverage, and skeleton feature vectors

Fig. 4.17 Clustering based on concatenated boundary, moment, coverage, and skeleton feature vectors

Fig. 4.18 Medoids from clustering based on concatenated boundary, coverage, and skeleton feature vectors

Fig. 4.19 Clustering based on concatenated boundary, coverage, and skeleton feature vectors

4.5 Discussion

Our measures capture most of the aspects of complexity outlined in Section 4.2—
except for the predictive approach which we did not explore here. Indeed, it appears
that all of these aspects of complexity are required to fully capture the range of shape

complexity: the most successful clustering appears to be the concatenated measures without moments shown in Figure 4.19. Interestingly, most of the medoids of that cluster are drawn from the symmetric shape class.

In particular, clustering based on boundary feature vectors (see Figure 4.11) is the only clustering that manages to group all mostly long, skinny shapes into the same cluster: watch, bone, knife, hammer, guitar, key, fork, shoe, and phone.

The skeleton feature vectors (see Figure 4.9) offer several convincing clusters but fail in two fundamental ways. One, visually simple shapes such as the rounded square and the heart are grouped with much more complex shapes. This is because the skeleton generates multiple branches to describe near-circular regions, making the skeletal description much more complex than the boundary. The skeleton measures also do not group the hammer in with the visually most similar shapes: bone, fork, and knife. On the other hand, the skeleton feature vectors produce the only clustering in which the deer, fly, and beetle, arguably the most complex shapes, all occur in the same grouping.

Area measures, shown in Figure 4.15, very successfully group together the large low-complexity objects, except for the rounded square in a cluster by itself. As the only shape that takes up almost the entire field, the clustering result makes sense. Other clusters, such as the one containing shapes with long, pointy parts, also seem appropriate. On the other hand, the complex beetle is in a cluster with the much simpler fork instead of in the cluster with the fly and deer.

Boundary measures, meanwhile, cluster the beetle and the deer together but group the fly with the less complex four-legged animals; see Figure 4.11. Boundary measures do, however, cluster simple shapes like the rounded square and hat together and also cluster symmetric shapes together despite having no explicit symmetry measure.

The global measures in Figure 4.7 cluster the symmetric shapes together as well, though they include the moment invariants, but also fail to cluster the high-complexity beetle, fly, and deer together. At first glance, the global feature vectors appear to cluster all of the simpler animals together, but the dog is grouped with flowers and with detailed petals. It may be that the moment invariants in some cases overwhelm the more geometric measures.

The combination of multiple feature vectors via concatenation seems to capture the best clustering characteristics of the individual feature vector sets. The moments appear to distract, however, as shown in Figure 4.17. The single cluster containing the spiral shape, the deer, and the grouping of some of the simplest shapes (rounded triangle and semi-circle) with some of the most complex (beetle and fly) are some of the most obvious inconsistencies.

Combined multiple feature vectors without the moment invariants (boundary, coverage, and normalized skeleton feature vectors) produce the most intuitively correct results; see Figure 4.19. Simpler shapes with long, slender protrusions (except for the deer) are grouped together, as are the simplest shapes such as heart and rounded square.

In short, each of the types of measures provides important information about a shape's complexity, and therefore the aggregate results are the most accurate.

Moreover, our results suggest that measuring complexity measures at multiple scales captures something important. Finally, we note that while clustering does not provide an ordering of shapes based on their complexity, we can attach complexity orders to the clusters based on membership in a cluster of a shape with a known complexity level.

4.5.1 Discussion of Methodology and Future Work

Our overall goal is to explore the qualitative effect of possible quantitative measures when those measures are no longer single numbers (and hence the shapes cannot be linearly sorted). Clustering by similarity allows us to visually examine the measures on actual shapes. We present six different combinations of measures here, each with six clusters in order to illustrate the approach.

There are many different ways to vary this approach: (1) use a broader range of shapes, (2) use nonlinear clustering methods (the Euclidean distance may not be adequate), and (3) use further analysis to determine which features are the most important in any given clustering. Further exploratory analysis of this kind could yield a "visual dictionary" that qualitatively defines the different measures and a better understanding of how to mathematically combine them to produce the desired groupings. Additionally, we have not yet analyzed which features are most significant in determining cluster membership.

It also seems clear that our choice of six clusters might not be ideal for each measure. We chose this number to ensure that the splitting is conservative—*i.e.,* we may have two clusters that are similar, but we will not have a cluster that should be split further. This is a possible explanation for why some shapes that seem similar (such as the devices) end up in two or more clusters. It should be noted that decreasing the number of clusters will not, in general, simply merge two clusters. Instead, we plan to explore the optimal number of clusters in each instance separately in future work.

Finally, we intend to implement an edit-based complexity measure to capture more explicitly the process displayed in Figure 4.2. We may also begin addressing probabilistic constructions that will allow us to explore the predictive approach to complexity.

Another key direction to explore is *supervised learning*, where we are provided with desired clusters (or relative rankings) and the goal is to find the combination of measures that best produce those clusters or rankings. The rich set of measures provided here (and their ability to produce different qualitative groupings) show promise for supporting this endeavor. For example, one direction we would like to explore is using user studies (such as the one in [2]) to produce labeled data and then using techniques such as rank support vector machines [6] to determine how to combine the measures in order to produce similar relative rankings. Note that this will provide the complexity ordering that our clustering method currently does not.

References

1. Bober, M.: MPEG-7 visual shape descriptors. IEEE Trans. Circuits Syst. Video Technol. **11**(6), 716–719 (2001)
2. Carlier, A., Leonard, K., Hahmann, S., Morin, G., Collins, M.: The 2D shape structure dataset: a user annotated open access database. Comput. Graph. **58**, 23–30 (2016)
3. Chazelle, B., Incerpi, J.: Triangulation and shape-complexity. ACM Trans. Graph. **3**(2), 135–152 (1984)
4. Chen, Y., Sundaram, H.: Estimating the complexity of 2D shapes. In: Proceedings of Multimedia Signal Processing Workshop (2005)
5. Feldman, J., Singh, M.: Information along contours and object boundaries. Psychol. Rev. **112**(1), 243–252 (2005)
6. Joachims, T.: Training linear SVMs in linear time. In: Proceedings of the 12th ACM SIGKDD International Conference on Knowledge Discovery and Data Mining, pp. 217–226. ACM, New York, NY (2006)
7. Koenderink, J.J., van Doorn, A.J.: Surface shape and curvature scales. Image Vis. Comput. **10**, 557–565 (1992)
8. Larsson, L.J., Morin, G., Begault, A., Chaine, R., Abiva, J., Hubert, E., Hurdal, M., Li, M., Paniagua, B., Tran, G., et al.: Identifying perceptually salient features on 2D shapes. In: Research in Shape Modeling, pp. 129–153. Springer, Cham (2015)
9. Leonard, K.: Efficient shape modeling: epsilon-entropy, adaptive coding, and boundary curves -vs- blum's medial axis. Int. J. Comput. Vis. **74**(2), 183–199 (2007)
10. Leonard, K., Morin, G., Hahmann, S., Carlier, A.: A 2D shape structure for decomposition and part similarity. In: International Conference on Pattern Recognition (2016)
11. Liu, L., Chambers, E.W., Letscher, D., Ju, T.: Extended grassfire transform on medial axes of 2D shapes. Comput. Aided Des. **43**(11), 1496–1505 (2011)
12. McCrae, J., Singh, K.: Sketching piecewise clothoid curves. In: Proceedings of the Fifth Eurographics Conference on Sketch-Based Interfaces and Modeling, pp. 1–8. Eurographics Association, Aire-la-Ville, Switzerland (2008)
13. Mercimek, M., Gulez, K., Mumcu, T.V.: Real object recognition using moment invariants. Sadhana **30**(6), 765–775 (2005)
14. Mitra, N.J., Wand, M., Zhang, H., Cohen-Or, D., Kim, V., Huang, Q.X.: Structure-aware shape processing. In: ACM SIGGRAPH 2014 Courses, pp. 13:1–13:21. ACM, New York, NY (2014)
15. Osada, R., Funkhouser, T., Chazelle, B., Dobkin, D.: Shape distributions. ACM Trans. Graph. **21**(4), 807–832 (2002)
16. Page, D.L., Koschan, A.F., Sukumar, S.R., Roui-Abidi, B., Abidi, M.A.: Shape analysis algorithm based on information theory. In: International Conference on Image Processing, pp. 229–232 (2003)
17. Panagiotakis, C., Argyros, A.: Parameter-free modelling of 2D shapes with ellipses. Pattern Recogn. **53**, 259–275 (2016)
18. Rigau, J., Feixas, M., Sbert, M.: Shape complexity based on mutual information. In: 2005 International Conference on Shape Modeling and Applications, 15–17 June 2005, Cambridge, MA, USA, pp. 357–362 (2005)
19. Schwarz, M., Wonka, P.: Practical grammar-based procedural modeling of architecture: Siggraph Asia 2015 course notes. In: SIGGRAPH Asia 2015 Courses, pp. 13:1–13:12. ACM, New York, NY (2015)
20. Sukumar, S., Page, D., Gribok, A., Koschan, A., Abidi, M.: Shape measure for identifying perceptually informative parts of 3D objects. In: Third International Symposium on 3D Data Processing, Visualization, and Transmission, pp. 679–686 (2006)

Chapter 5
Phase Field Topology Constraints

Rüyam Acar and Necati Sağırlı

Abstract This paper presents a morphological approach to extract topologically critical regions in phase field models. There are a few studies regarding topological properties of phase fields. One line of work related to our problem addresses constrained phase field evolution. This approach is based on modifying the optimization problem to limit connectedness of the interface. However, this approach results in a complex optimization problem, and it provides nonlocal control. We adapted a non-simple point concept from digital topology to local regions using structuring masks. These regions can be used to constrain the evolution locally. Besides this approach is flexible as it allows the design of structuring elements. Such a study to define topological structures specific to phase field dynamics has not been done to our knowledge.

5.1 Introduction

In this paper, we address the problem of detecting topologically critical regions in phase field modeling. Extracting topological properties of digital data can be studied in two different approaches. The first approach is based on embedding the continuous topological structures in a domain discretized into simplicial cells. In this context, discrete Morse theory, Reeb graphs, and Morse-Smale complex can be considered to analyze topological properties of the data. In the second approach, topological properties are defined on a digital grid based on graph and set theories. In this case, topological concepts derived from these definitions, such as simple points, can be used. Topological information obtained from digital data is essential to various applications such as visualization, shape analysis, and skeletonization. Constraining the topological changes in image transformations or deformations is also one of the important problems. Requirements of topological

R. Acar (✉) · N. Sağırlı
Okan University, Istanbul, Turkey
e-mail: ruyam.acar@okan.edu.tr; necati.sagirli@hotmail.com

© The Author(s) and the Association for Women in Mathematics 2018

A. Genctav et al. (eds.), *Research in Shape Analysis*, Association for Women in Mathematics Series 12, https://doi.org/10.1007/978-3-319-77066-6_5

feature extraction also depend on the application and data representation. While Morse-Smale complex is more suitable for visualization purposes, digital topology concepts could be more advantageous in constraining transformations.

In implicit modeling, mostly constrained deformation problem is addressed. In level sets, non-simple point concept from digital topology is used to constrain topological evolution. Note that using critical points of Morse functions can also be considered for level set modeling. In phase field modeling, topology constraints are embedded in energy optimization [18]; distance-based connectedness constraint is included in the energy function to control the evolution of the interface. However, this results in a complex optimization problem, and more importantly this formulation provides only nonlocal controls. Simple point concept can also be used in phase field modeling to obtain local control. However, phase field dynamics involves diffuse topological regions. Besides, using singular points in phase field equations would cause instabilities. For these reasons, we aimed at defining critical regions rather than single points. In this paper, we designed simple morphological masks to adapt the non-simple point concept to critical regions. These masks force the connected component counts to be in the vicinity of opposing surfaces and in accordance with phase field topological transitions. Currently, we obtain nondiffuse critical regions and with some error tolerance due to the limitations of the imposed directions. However in the future we would like to extend these masks to adaptive morphological operators with geodesic distance values to obtain more accurate and diffuse regions.

In the next section, we will give a brief review of main concepts from topology theory. Then discrete topology approaches will be reviewed. Next, we will discuss related methods, specifically topology studies in the implicit framework, namely, level sets and phase fields. In the following section, we will discuss topological requirements of phase field models and describe our methodology. Finally, we will discuss future extensions and improvements to our method.

5.2 Topology

General topology is a well-established theory; it involves the study of structural properties of space such as connectedness, compactness, and continuity. The emphasis is especially on the preservation of these properties under continuous deformations. These properties are studied through set theory and geometry. *Algebraic topology* studies classification of topological spaces through algebraic tools such as homology and homotopy. Homotopy classifies topological properties based on their invariance under continuous deformation. Two spaces X and Y are homotopic if one can be transformed into the other continuously. Homology group is related to the number of holes in a topological space. Other important concepts from algebraic topology are Euler characteristics and Jordan curve theorem. Euler characteristics relate the number of edges, vertices, and faces in an object to its topological properties. It is a global topological invariant. Jordan curve theorem states that a closed curve separates a plane into two regions.

Morse theory studies critical points; this theory deals with the topology of a surface by investigating the critical points of a smooth (differentiable) function on the surface. Critical points are defined as the minima, maxima, and saddles of this function, and they are isolated. Reeb graph, which is also a powerful tool in topological analysis, builds a topological map of space using the connectivity of the level set components and the critical points of Morse function. However, Reeb graphs may not be able to provide all topological information; for example, some of the cavities may be missed since the graph structure does not represent the geometric connectivity of features completely. Morse-Smale complex builds a more complete topological structure based on an analysis of the gradient flow of the data. Here, instead of the level set contours, the gradient flow behavior between critical points is used. In order to represent this flow behavior, integral lines connecting the critical points are built. This process decomposes the image into smooth regions separated by integral lines, and it provides a more robust and structurally stable topological description.

5.2.1 Discrete Topology

Topological analysis of discrete fields can be approached in two ways. In the first approach, space is decomposed into simplicial cells, and continuous topological properties are embedded in discrete space. This is called *combinatorial topology*. The second approach, so-called digital topology, defines topological concepts such as adjacency and continuity on a digital grid based on set and graph theories. Other topological properties are then derived based on these definitions. Both approaches are relevant to the topological analysis of implicit surfaces such as level sets and phase fields. We will first discuss the combinatorial approach for Morse theory very briefly. Then we will give a more detailed review of digital topology since our method is based on this approach. However, it should be mentioned that discretization of topological properties is not a trivial matter; adaptation of continuous properties on discrete fields might cause information loss.

Discrete Morse theory is based on the combinatorial approach; discrete Morse functions are defined on simplicial complexes. Other methods based on Morse theory such as Reeb graphs and Morse-Smale complex use these discrete Morse functions as the basis of their computation. However, discrete Morse theory suffers from stability issues, and computation of Morse-Smale functions can be problematic due to the sensitivity to noise [7].

5.2.2 Digital Topology

The foundation of digital topology lies in the union of open sets [10]. The basic notion of digital topology, connectivity, is derived from a graph theoretic approach.

Accordingly adjacency relations of a point are defined based on the neighborhood of a point on a cellular grid. Although various adjacencies are defined, the most common connectivity definition is based on the 4 and 8 neighbors of a point. Defined by Rosenfeld [11], these are known as 4- and 8-adjacencies. Connected components obtained based on these adjacencies form the homotopy groups in digital topology. Once the connectivity relations are defined, the main topological concepts such as homology, fundamental group, isomorphism, and path can be defined based on these relations. Fundamental group is the simplest homotopy group; it represents the continuous deformation of one path to the other [13]. Homeomorphism indicates spaces with the same homotopy groups; homeomorphic spaces are considered topologically equivalent. In Rosenfeld digitization (which is based on 4- and 8-adjacencies) Jordan curve theorem cannot be reproduced [9, 14]. For a curve to separate a plane into two components, the curve and the plane should have different adjacencies in this digitization. Khalimsky [9] proposed different adjacency relations based on connected ordered topological spaces; Jordan curve theorem can be derived based on these connectivities.

Combinatorial approach, that is embedding of continuous topological structures into the digital grid, provides a basis for the derivation of some theorems such as Euler theorem and allows the adaptation of important topological concepts such as Euler characteristic and homology groups [5, 10]. Another important concept in digital topology is *simple points*. In fact, the notion of simple points together with the connectivity relations forms the basis of digital topology. A simple point [11] is defined as a point whose deletion does not change the number of components in the image (Figure 5.1). Global topological properties of the image are related to the number of components on the grid. Basically, topology is preserved in a space if the number of components is preserved. Consequently, digital homotopy states that a space X can be obtained from space Y through continuous insertion of simple points. Also points which are not simple are considered as critical points.

Various different concepts are defined in digital topology both due to the ambiguities or deficiencies in digitization or due to the requirements of the specific application. For example, a simple set [16] is defined as a set of points whose removal does not alter the topology of space. In other words, instead of removing a single point each time, the effect of simultaneous removal of a group of points is considered. Accordingly critical regions or minimal sets (simple sets with a minimum size) are defined [17]. These properties are defined for thinning algorithms [2].

5.3 Discussion

Although the basics of topology theory are well founded, it has been studied for decades since there are still several open problems and still new connections are being discovered. Discretization of topological structures or adaptation of topological concepts in the discrete domain brings out a new list of open problems.

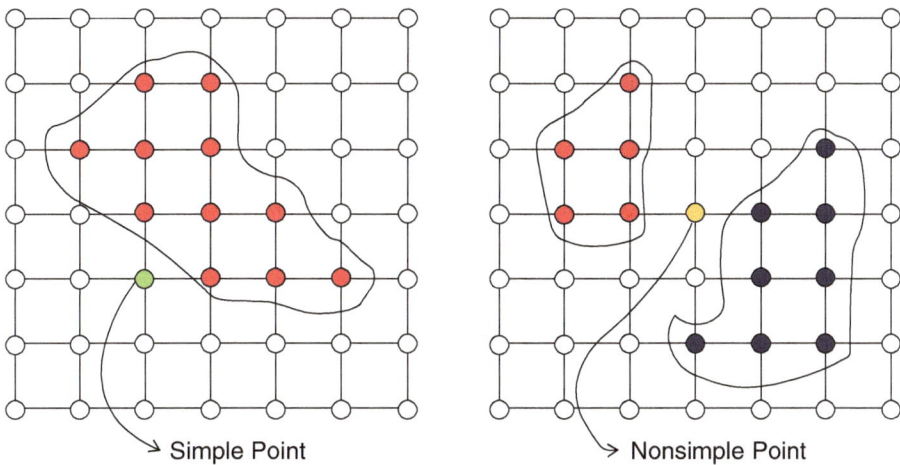

Fig. 5.1 Simple and non-simple points in digital topology based on 8-adjacencies

Another factor that affects the course of studies in the discrete domain is the variety of applications. Several problems that arise in discrete topology are application specific. Topological information is essential to various problems in image processing and shape analysis such as shape description and feature extraction, visualization, thinning, segmentation, registration, and reconstruction. Some of these applications require the control of topological properties of an image which undergoes some processes or transformations. For shape description or visualization, the use of Reeb graphs or Morse-Smale complex is advantageous since these methods provide a topological map of the whole image. Reeb graphs are also useful in building level set functions and skeletonization. Recently Ngo et al. [15] used an adjacency tree built upon the adjacency relations of connected components to preserve the global topology of an image which goes through rigid transformations. However, some applications such as nonrigid transformations require local constraints, and usually simple point idea is used for this purpose. Faisan et al. [6] define image deformation as a constraint-based optimization. They define a cost function based on a one-to-one mapping between the connected components, and this mapping is estimated from the movement of simple points in the direction of deformation. In this way, the control of addition and removal of simple points are embedded in the cost function.

In the level set framework, also simple point concept from digital topology is commonly used as a local constraint. The sign change in the distance function indicates a region with surface change; the sign of distance function changes when a surface is deformed due to level set evolution. This corresponds to the insertion and deletion of points from the surface in digital topology. Therefore critical points should be in these regions. Han et al. [8] combine this property with the simple point concept to find the critical points in the level set evolution. In regions where a sign change takes place, they look for non-simple points whose removal or insertion changes the number of components in a local neighborhood. Note that they apply thresholding on the distance function to obtain the binary image. They also compute Euler characteristics from the contour evolution; Euler number is then used to verify topological properties globally. Later Alexandrov et al. [1], and Guyader et al. [12] proposed a topological constraint based on the geometric properties of the level set function. Using concepts from Morse theory might also be considered to detect topological properties in level set evolution. However critical points of the Morse function are isolated, and computation can be problematic as discussed previously.

In phase field modeling, topology preservation is incorporated in energy minimization. Dondl et al. [3] include a penalty term in Willmore energy to control surface connectedness. The penalty term uses a geodesic distance function to control the length of connected components. Based on connectedness, this is a nonlocal term. Due to the distance function calculations, computation of the penalty term and the solution of the optimization problem can be expensive. Topological events in phase field evolution are local processes. Wojtowytsch [18] proposed using a blow-up control in the energy function and derived a relaxed Willmore energy with a local control parameter to limit topological evolution. However, computability of this equation has not been proven yet. Another line of work in phase field modeling focuses on the computation of topological numbers. Du et al. [4] derived an integral formula based on the phase field values to compute Euler number on phase field interfaces. However, Euler number, encompassing a global information, cannot be used to control topological events.

5.4 Critical Region Detection in Phase Field Modeling

Identification of topologically critical regions in phase field surfaces is necessary in several applications especially for constraining the topological evolution. Incorporating this problem in energy optimization as suggested in [18] results in a nonlocal and computationally expensive control mechanism. Simple point concept from digital topology can also be used here as in the level set model. In this case, regions that are prone to topological changes need to be detected first. In the level set method, this is achieved by tracking the sign change of the level set function. Once these areas are detected, at each point simple point criteria can be tested in a local neighborhood. However, the phase field surface has a transition region, and critical areas are also transitional regions rather than single points. Besides, using only single points in the phase field equations as a limiting parameter will cause instability. Thus, critical areas should be determined in accordance with the properties of both the surface representation and phase evolution in phase field dynamics.

In phase field dynamics, when two surfaces become close, the transition layer formed between them triggers an energy transfer between the surfaces. The width of this transition layer depends on the surface thickness. Phase values in this transition region are attracted toward one of the states due to the energy flow. As a result, coalescence occurs. Breakup event happens similarly, only for the background phase

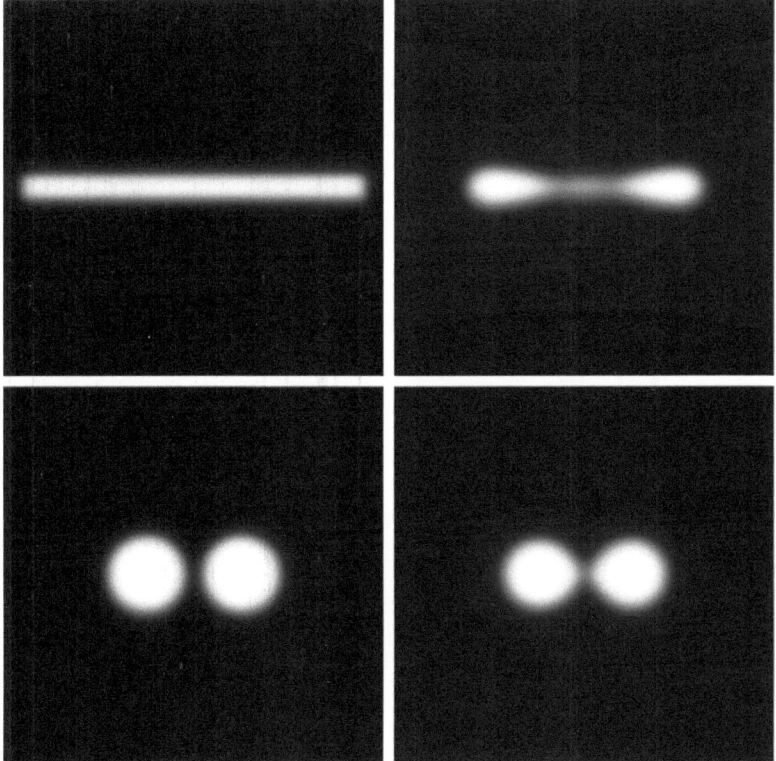

Fig. 5.2 Transitional regions and layer formation in breakup (*top*) and coalescence (*bottom*) events. Initial phase values (*left*) and a number of steps after phase field evolution (*right*)

values (Figure 5.2). The critical regions should comply with this layer formation both geometrically and in terms of phase values. Currently, we use the following approach to detect critical regions in a phase field model.

First, we obtain a binary image by thresholding the phase values. We extend the non-simple point idea from digital topology to a regional description using structure masks. Generally speaking, a critical (non-simple) region is a set of points in the vicinity of two components on opposite sides. If this is a background (foreground) region, then the opposing components should be foreground (background) components. To obtain the geometric constraint enforced by opposing directions, we use masks in four principal directions as shown in Figure 5.3. In the beginning, connected components are labeled globally using a two-pass algorithm. 8-adjacencies are used for the connectivities. Then at each point, four masks are applied separately and in a local window. In each mask, the center points are assigned a 0 value and the rest are 1. This is to separate the critical region from the two opposing regions. Then the component labels in regions with a value 1 are counted. If there is at least one component in each opposing region, then the central

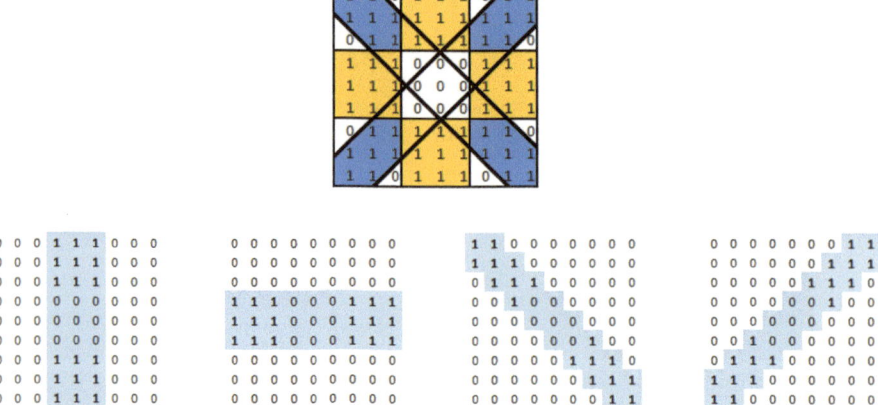

Fig. 5.3 Structuring masks in four directions

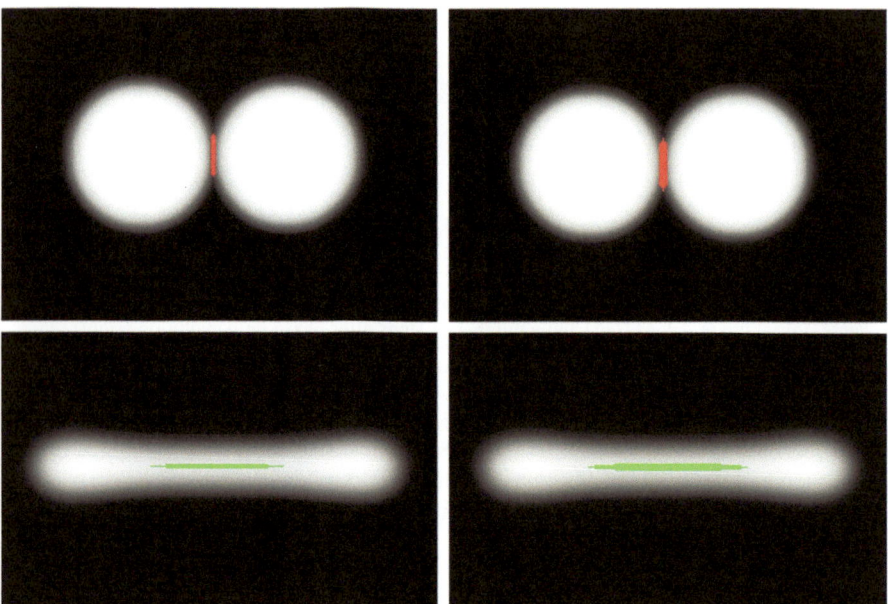

Fig. 5.4 Critical region detection with two different window sizes. Coalescence (*top*) and breakup (*bottom*). Window size = 9 × 9 (*left*) and window size = 11 × 11 (*right*)

area is considered as a critical region. Note that, here, non-simple point criteria are adapted locally. That is, the change in the number of components is checked locally. The size of the local window is adjusted according to the surface thickness. In this way, we obtain the critical, transitional regions between surfaces (Figure 5.4). Note that masks are designed to be slightly overlapping to cover all the points in

Fig. 5.5 Critical regions detected with the proposed approach during phase separation. Window size $= 5 \times 5$

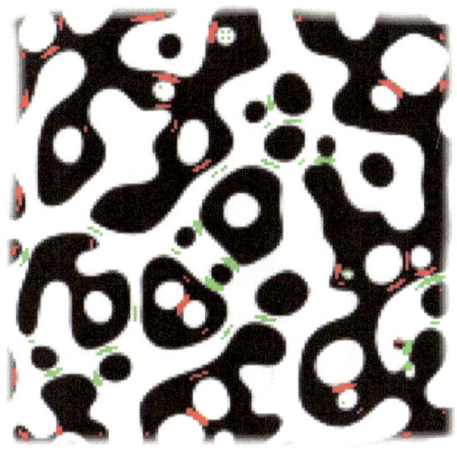

a window. Figure 5.4 shows the critical regions obtained with this approach for different window sizes. Figure 5.4 shows critical regions for an image with different surface properties. (This image shows a frame from phase field evolution.)

5.5 Future Work

We have presented a morphological approach to determine topologically critical regions in phase field dynamics. We do not find the non-simple point criteria from digital topology appropriate for this purpose mainly for two reasons. First, if the goal is to use the critical points to constrain phase field evolution, then using isolated points as a parameter in the equation will cause instabilities in the numerical solution. Second, in phase field dynamics, topologically critical areas are in fact diffuse regions.

In this work, we use structuring masks to adapt the non-simple point idea to regions. Masks are designed to force the basic geometric properties of phase field surfaces in critical areas. With this method, we obtained nondiffuse critical regions in phase field models although somewhat roughly. Our goal is to extend these rather simple masks to adaptive structuring elements which include geodesic distance values to obtain diffuse critical regions with the same geometric properties as in phase field dynamics. Using these diffuse regions in phase field equations to obtain topology constrained phase field evolution will be our next task. Constrained phase field evolution has various potential applications in both computational physics and graphics.

Another line of work that we would like to focus on is improving the computational efficiency of our algorithm. Currently, masks are used to update single pixels. However, updating and labeling regions rather than only one pixel at a time would improve the efficiency significantly. Also, in constraining the phase field evolution, updating the critical regions at each step, would slow down the simulation. To solve this problem, using the similarities in the flow of critical regions can be considered. Finally, it would be interesting to explore extensions of these morphological operators to obtain nonlocal shape features.

References

1. Alexandrov, O., Santosa, F.: A topology-preserving level set method for shape optimization. J. Comput. Phys. **204**(1), 121–130 (2005)
2. Bertrand, G., Couprie, M.: On parallel thinning algorithms: minimal non-simple sets, p-simple points and critical kernels. J. Math. Imaging Vis. **35**(1), 23–35 (2009)
3. Dondl, P.W., Lemenant, A., Wojtowytsch, S.: Phase field models for thin elastic structures with topological constraint. Arch. Ration. Mech. Anal. **223**(2), 693–736 (2015)
4. Du, Q., Liu, C., Wang, X.: Retrieving topological information for phase field models. SIAM J. Appl. Math. **65**(6), 1913–1932 (2005)
5. Eckhardt, U., Latecki, L.: Digital Topology. In Current Topics in Pattern Recognition Research, Research Trends, Council of Scientific Information, Vilayil Gardens, Trivandrum (1995)
6. Faisan, S., Passat, N., Noblet, V., Chabrier, R., Meyer, C.: Topology preserving warping of 3-d binary images according to continuous one-to-one mappings. IEEE Trans. Image Process. **20**(8), 2135–2145 (2011)
7. Gunther, D.: Topological analysis of discrete scalar data. Ph.D. thesis, Max-Planck-Institut Informatik (2012)
8. Han, X., Xu, C., Prince, J.: A topology preserving level set method for geometric deformable models. IEEE Trans. Pattern Anal. Mach. Intell. **25**(6), 755–768 (2003)
9. Khalimsky, E.: Topological structures in computer science. J. Appl. Math. Simul. **1**(1), 25–40 (1987)
10. Klette, R., Rosenfeld, A.: Digital Geometry: Geometric Methods for Digital Picture Analysis. Morgan Kaufmann, San Francisco (2004)
11. Kong, T., Rosenfeld, A.: Digital topology: introduction and survey. Comput. Vis. Graph. Image Process. **48**(3), 357–393 (1989)
12. Le Guyader, C., Vese, L.A.: Self-repelling snakes for topology-preserving segmentation models. IEEE Trans. Image Process. **17**(5), 767–779 (2008)
13. Malgouyres, R.: Presentation of the fundamental group in digital surfaces. In: Bertrand, G., Couprie, M., Perroton, L. (eds.) Discrete Geometry for Computer Imagery. DGCI 1999. Lecture Notes in Computer Science, vol. 1568, pp. 136–150. Springer, Berlin (1999)
14. Melin, E.: Connectedness and continuity in digital spaces with the Khalimsky topology (2003)
15. Ngo, P., Passat, N., Kenmochi, Y., Talbot, H.: Topology-preserving rigid transformation of 2D digital images. IEEE Trans. Image Process. **23**(2), 885–897 (2014)
16. Passat, N., Mazo, L.: An introduction to simple sets. Pattern Recogn. Lett. **30**(15), 1366–1377 (2009)
17. Ronse, C.: Minimal test patterns for connectivity preservation in parallel thinning algorithms for binary digital images. Discret. Appl. Math. **21**(1), 67–79 (1988)
18. Wojtowytsch, S.: Phase-field models for thin elastic structures: Willmore's energy and topological constraints. Ph.D. thesis, Durham University (2017)

Chapter 6
Adaptive Deflation Stopped by Barrier Structure for Equating Shape Topologies Under Topological Noise

Asli Genctav and Sibel Tari

Abstract Using level sets of a pair of transformations, we adaptively bring two shapes to be matched to a comparable topology prior to a correspondence search. One of the transformations readily provides a central structure for each shape. We utilize the central structure as a reference volume for scale normalization. By adaptively dilating the central structure with the help of the second transformation, we construct what we refer to as the *barrier* structure. The barrier structure is used to automatically stop topology equating adaptive deflations. Illustrative experiments using different datasets demonstrate that our approach provides robust solutions for the topological noise caused by localized touches or spurious links that connect different shape parts.

6.1 Introduction

Advancement of visual data acquisition technology enabled easy acquisition of 3D shapes in the form of surface meshes enclosing solid objects. These meshes are used in many computer vision and graphics applications, many of which require establishing meaningful correspondences that pair up semantically equivalent points on two surfaces. The process of pairing up is called *matching*. For a matching result to be of practical value, the matched points should be semantically equivalent where the semantic equivalence needs to be inferred from the geometrical and topological properties. Typically, however, geometrical and topological information is corrupted by noise, which may get added either during acquisition or model formation.

The methods that address the problem of matching under topological noise can be classified into two groups: model-based and model-free. The techniques in the first group require a prior shape model. The topological complications are resolved by aligning the prior model with each one of the shapes [17]. A disadvantage of model-

A. Genctav (✉) · S. Tari
Department of Computer Engineering, Middle East Technical University, Ankara, Turkey
e-mail: asli@ceng.metu.edu.tr; stari@metu.edu.tr

© The Author(s) and the Association for Women in Mathematics 2018
A. Genctav et al. (eds.), *Research in Shape Analysis*, Association for Women in Mathematics Series 12, https://doi.org/10.1007/978-3-319-77066-6_6

based techniques is that they require good prior models which may not always be available. In such situations, a model-free approach may be the only option.

One model-free technique is to register a pair of 3D shapes via their spectral embedding, after eigenvector reordering [9]. Despite the effort spent on reordering, the matching is tolerant only to moderate noise. The most common strategy in model-free techniques is to replace topology-sensitive distances with robust ones [1, 2, 10–12]. Note that a measure of dissimilarity (distance) is required if one wants to find the similar pairs. The usual distance is the geodesic distance (which depends on the shortest path between a pair of points). Naturally, the choice of the distance affects the quality of the matching. For example, the geodesic distance is robust only to certain class of deformations, yet sensitive to topological changes. Diffusion-based distances corresponding to an average of all the paths between the pair of points are less sensitive, hence, commonly employed in the literature [1, 2, 10–12]. The diffusion distances, however, are sensitive to the choice of the scope of averaging, *i.e.*, the scale parameter. Therefore, in some works [10, 12], diffusion distances at multiple scales are employed to achieve robustness to topological changes.

In our work, we focus on a particular type of topological noise: one that is unavoidable even if the physical capabilities of the acquisition system are of very high quality, e.g., an arm touching the body during a motion capture session (Figure 6.1). Our approach is to adaptively deflate the pair of shapes for the purpose of bringing the pair of shapes to be matched to a comparable topology before the search for the correspondences. Once the pair of shapes is brought to a topologically comparable form, the matching is performed between topologically comparable forms, and then the found correspondences are transferred to the correspondences between the original pair of input shapes. Previously, Genctav et al. [6] inflated and deflated surface meshes with the help of a smooth indicator function proposed in Tari et al. [16]. In the present article, we propose a new and more systematic

Fig. 6.1 A pair of 3D shapes from MIT samba sequence [17]. Model on the right has hands connected to the belly

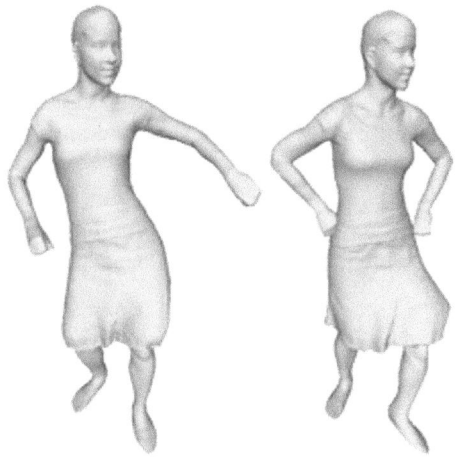

topology equating process by employing two volumetric transformations, each of which is an approximation of the Euclidean distance transform (EDT) subject to competing regularities. This pair of transformations enables us to implement the deflation process more systematically, to robustly handle scale normalization, and to define a stopping condition. We denote the transformations by ρ^0 and ρ^1 where the superscripts 0 and 1 indicate the values of a binary parameter used in their computation via a common model presented in the next section.

6.2 Adaptive Deflations via ρ^0 and ρ^1

The transformation ρ^0 is merely a smooth approximation of the EDT. It facilitates the generation of a collection of iso-surfaces that represent adaptive shape simplifying deflations of the shape boundary parameterized by a deflation level $\ell \in [0, 1]$; the larger the ℓ, the more is the deflation, hence the simplification. The deflations are adaptive: At any fixed level $\ell \in [0, 1]$, the amount of deflation at a point p of the deflating surface depends on the surface features that are implicitly coded through the values of ρ^0 in an $\epsilon-$neighborhood of p. We discretize ℓ by sampling its range $[0, 1]$ at a fixed length $\delta = 0.004$. We determine δ empirically where the number of iso-surfaces generated for $\delta = 0.004$ provided enough resolution in our experiments. Automatic determination of δ is the subject of a future study.

Deflations are merely selected iso-surfaces of ρ^0. In order to make sure that the process of deflating stops at an appropriate level so that the set of deflations is of practical value, *i.e.*, the process does not yield trivial surfaces for any $\ell \in [0, 1]$, we construct a *barrier structure*. The barrier structure is used to automatically stop topology equating adaptive deflations. The details of the construction process which makes use of ρ^1 and the EDT will be given in a later subsection.

Once we have a collection of iso-surfaces for each of the shapes, we follow the same steps as in Genctav et al. [6] in order to choose the pair of iso-surfaces that is to be considered as topologically comparable forms of the input shapes. First, we determine the levels at which the iso-surfaces from both shapes have comparable topology. We quantify topology of the iso-surfaces in terms of their genus number that intuitively counts the number of handles of a given object. In order to compute genus number of the iso-surfaces, we employ the Euler formula which is applicable to closed and connected manifold meshes. Suitably, the extracted iso-surfaces are always composed of closed meshes since the corresponding transformation ρ^0 is smooth and continuous. However, the iso-surfaces can have more than one mesh component, and each component may contain non-manifold vertices or edges. Thus, we consider the largest mesh component in terms of vertex count and apply the Euler formula after checking its manifoldness. Second, we take the smallest one among the genus numbers shared between pairs of iso-surfaces. The iso-surfaces are more similar to input shapes at preceding levels so we choose the first pair of iso-surfaces corresponding to the selected genus as the topologically comparable forms of the input shapes.

6.2.1 Computing ρ^β

We now explain how ρ^0 and ρ^1 are computed via a common formulation. Let $\Omega \subset Z^3$ denote the volume enclosed by the surface mesh of the shape where Z is the set of integer numbers. Further let $\partial\Omega$ denote the boundary of Ω, i.e., a grid representation for the surface mesh. The two regularized approximations to the EDT, namely, ρ^0 and ρ^1, are obtained as the minimizers of the cost function given in (6.1), where the binary valued parameter $\beta \in \{0, 1\}$ is set to 0 and 1, respectively.

$$\sum_{(i,j,k)\in\Omega} E_d(\rho^\beta_{i,j,k}) + E_s(\rho^\beta_{i,j,k}) + \beta E_a(\rho^\beta_{i,j,k}) \tag{6.1}$$

$$E_d(\rho^\beta_{i,j,k}) = \frac{1}{O(|\Omega|)}\,(\rho^\beta_{i,j,k} - EDT^\Omega_{i,j,k})^2$$

$$E_s(\rho^\beta_{i,j,k}) = \Big[(\rho^\beta_{i+1,j,k} - \rho^\beta_{i-1,j,k})^2 + (\rho^\beta_{i,j+1,k} - \rho^\beta_{i,j-1,k})^2$$

$$+ (\rho^\beta_{i,j,k+1} - \rho^\beta_{i,j,k-1})^2\Big]/4$$

$$E_a(\rho^\beta_{i,j,k}) = \frac{1}{O(|\Omega|)}\,\Big(\sum_{(i',j',k')\in\Omega} \rho^\beta_{i',j',k'}\Big)^2$$

subject to $\rho^\beta = 0$ on $\partial\Omega$.

Notice that E_d decreases as ρ^β gets closer to the EDT, E_s decreases as ρ^β becomes smoother, and E_a decreases as the global average of ρ^β approaches 0.

The difference between the volumetric transformation ρ^0 and the smooth distance function used in the preliminary work [6] is the following. Roughly speaking, ρ^0 is a smoothing of the EDT, whereas the former one is a smoothing of the characteristic function which is 1 on the shape and 0 elsewhere. This minor distinction is indeed important in terms of obtaining a simplified adaptive level separation. In the following subsections, we show that the topology equating process takes less number of steps when we use ρ^0 to extract shape deflations.

The second volumetric transformation, ρ^1, is based on the $2D$ region segmentation field model from [14, 15]. ρ^1 is a smooth function that takes both positive and negative values so that the absolute value of its global average is minimized. ρ^1 is positive at the inner shape regions where the EDT is high, whereas it is negative at the outer shape regions where the EDT is low. The zero-crossing of ρ^1 provides a partitioning of the shape into central and the remaining outer regions. In a later work [5], this part coding behavior is obtained via local linear computations. In our work, we utilize ρ^1 to perform the scale normalization between the input pair of shapes as well as to extract a suitable stopping condition via what we refer as the barrier structure.

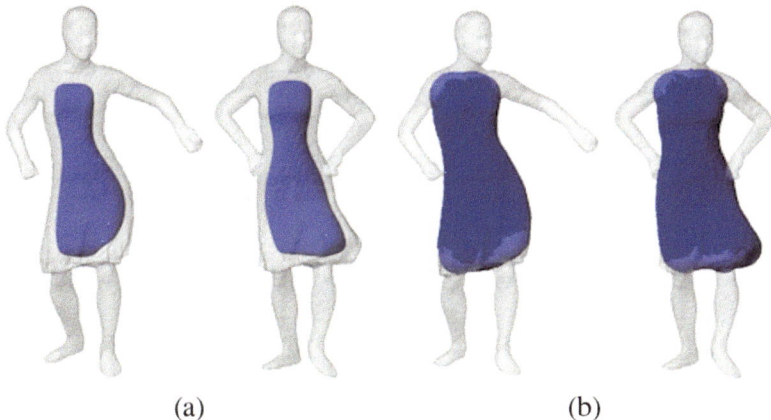

Fig. 6.2   (a) Central structures (b) Barrier structures for the shapes in Figure 6.1

6.2.2   The Barrier Structure

The volumetric transformation ρ^1 has two phases – negative (outside) and positive (inside). The phase boundary given by the zero-crossing of ρ^1 separates the entire volume (enclosed by a surface) into two disjoint sets, one of which captures the central structure as illustrated in Figure 6.2 (a). We dilate this central structure so that it touches the shape boundary and obtain the barrier structure, a coarse shape without appendages (see Figure 6.2 (b)). In order to extract the iso-surface that represents the maximum deflation allowed (the iso-surface at $\ell = 1$), we use $\rho^0_\star$ which is the maximum value of ρ^0 inside the shape volume and outside the barrier structure. Accordingly, the iso-surface at an intermediate level ℓ is extracted using the value $\ell \times \rho^0_\star$. In this way, we consider meaningful deflations preserving the essential shape structure. Notice that the deflations cannot fully enclose the barrier structure since the deflations are inside the shape boundary, whereas the barrier structure touches the boundary. Instead of stopping the deflations according to their intersection with the barrier structure which needs to be computed for each of them, we simply consider the iso-surface of ρ^0 with the value $\rho^0_\star$ as the maximum deflation.

6.2.3   An Illustration

In Figure 6.3, we present a shape from MIT samba sequence [17] (which is the second shape in Figure 6.1) along with the iso-surfaces of the corresponding ρ^0 at $\ell = 0.044$ and $\ell = 0.052$. Note that the hands of the shape are touching the belly, and the wrists disappear first in the deflated forms of the shape boundary because they are the thinnest parts of the shape. At $\ell = 0.052$, both of the arms are separated from the belly, while the essential shape structure is preserved.

Fig. 6.3 The second shape
in Figure 6.1 and iso-surfaces
of the corresponding ρ^0 at
$\ell = 0.044$ and $\ell = 0.052$

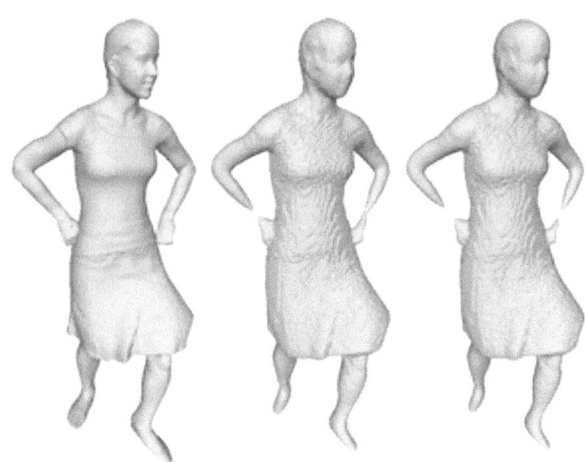

Table 6.1 Summary of the notions used throughout the text

Shape volume	$\Omega' \subset R^3$
Negative phase	$\{x \mid x \in \Omega' \text{ and } \rho^1(x) < 0\}$
Positive phase	$\{x \mid x \in \Omega' \text{ and } \rho^1(x) > 0\}$
Zero-crossing (phase boundary)	$\{x \mid x \in \Omega' \text{ and } \rho^1(x) = 0\}$
Barrier structure	Dilation of the positive phase so that it touches the shape boundary
$\rho_\star^0$	$\max(\{\rho^0(x) \mid x \in \Omega' \text{ and } x \notin \text{Barrier structure}\})$
Deflation at level ℓ where $\ell \in [0, 1]$	$\{x \mid x \in \Omega' \text{ and } \rho^0(x) = \ell \times \rho_\star^0\}$
Maximum deflation	Deflation at level $\ell = 1$

In Table 6.1, we present a summary of the notions used throughout the text. In Figure 6.4, we illustrate the topology equating process.

6.2.4 Comparison to the Preliminary Work [6]

We compare ρ^0 with the smooth function used in Genctav et al. [6] in terms of the number of deflation steps needed for obtaining topologically comparable forms of the input shapes. We use a set of 99 topologically different pairs of shapes from SHREC11 robustness benchmark [3]. The deflation at step k is obtained as the iso-surface of the corresponding function at level $0.004 \times k$ where each function is normalized to have the maximum value of 1. The average number of deflation steps is around 2 ± 7.7 using ρ^0, whereas it is around 6 ± 14.3 using the function in Genctav et al. [6]. We find that the topology equating process takes less number of steps when we use ρ^0 to extract shape deflations.

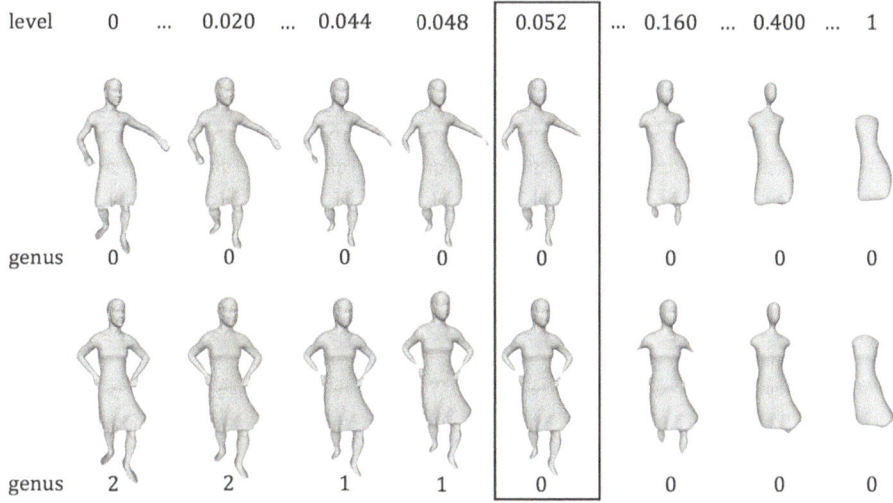

Fig. 6.4 Topology equating process involves extracting a collection of deflations for each shape and choosing a level at which the deflations have comparable topology. For this example, the selected level is 0.052 at which the genus number becomes the same for the deflations of both shapes

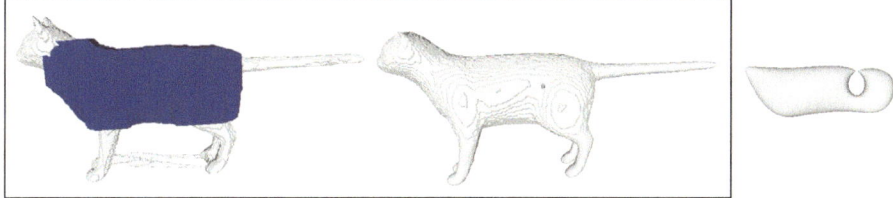

Fig. 6.5 (Left) Barrier structure for a cat shape with topological noise. (Middle) Deflation selected by our proposed topology equating strategy which uses the barrier structure. (Right) Deflation selected by Genctav et al. [6]

The advantage of restricting the collection of deflations via the barrier structure can be illustrated via the following example. Consider a pair of cat models each with links between different shape regions as well as a hole in the main body close to the hip section. The links are removed in early stages of the deflation process, whereas the hole is retained until very late stages. Accordingly, the genus number becomes 1 after a few steps, but it reduces to 0 when the deflation process breaks apart the hole meaning that the essential structures such as the head, legs, and tail are already removed. By utilizing the barrier structure in Figure 6.5 (left), our proposed topology equating strategy stops the deflation process before the removal of the hole and, hence, utilizes the pair of deflations with genus number 1 at the first smallest level (see Figure 6.5 (middle)). In Genctav et al. [6], the collection of deflations is not restricted so it includes all the deflations until the whole shape vanishes.

Fig. 6.6 A pair of shapes
from i3DPost Multi-View
Human Action dataset [7]
where 3D models are
reconstructed using [13]

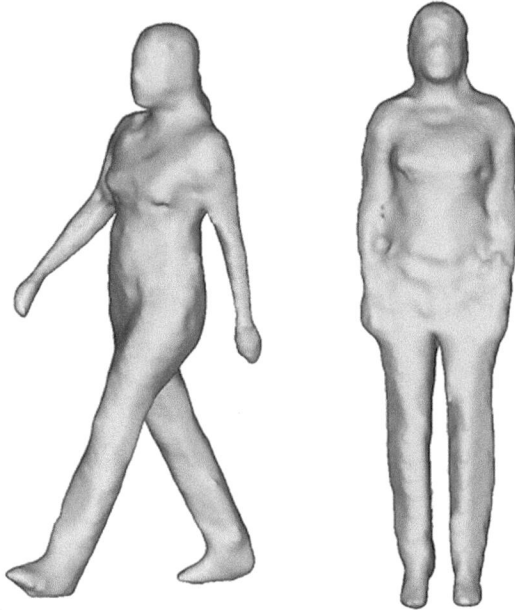

Accordingly, the selected pair of deflations with the smallest genus number becomes
the core regions of the shapes with genus number 0 after the removal of the holes
(see Figure 6.5 (right)).

We use the central structure for scale normalization. It gives more reliable
normalization as compared to geodesic distances which may be misleading in the
case of topological deformations. For example, consider a pair of human shapes
one of which has arms fully touching the body (see Figure 6.6). In this case, the
respective maximum geodesic distances for the two shapes are not semantically
equivalent. For the left shape, it is from hand to foot. For the right shape, it is from
head to foot. Hence normalization with respect to maximum geodesic distance is
not appropriate.

6.3 Experimental Evaluation

In order to evaluate our proposed method that brings a pair of shapes into
topologically comparable form via deflating the shape boundary, we use it in a
matching task. We transfer the mapping computed between the selected pair of
deflations to the input shapes by finding the vertices closest to the matched points
of the deflations.

We experiment with the following datasets: the samba sequence from MIT [17],
SHREC10 correspondence [4] and SHREC11 robustness [3] benchmarks, and the
flashkick sequence [13] from University of Surrey.

The samba sequence contains sequences of a dancing woman captured using a multi-camera system. Topological differences arise when the arms touch the body during motion. We illustrate our method on a topologically different shape pair from the sequence.

SHREC10 correspondence benchmark includes three objects. Each object comes with a base shape model and five additional forms obtained via isometric deformation of the base shape further deformed by adding topological noise of increasing strength. In our experiments, we match each base shape to the remaining five.

SHREC11 robustness benchmark contains 12 different shape models. For each model, there is one base shape, one isometric shape, and five shapes with increasing degree of topological noise. We use a subset for which the ground truth correspondence is available.

The flashkick sequence of a dancing man contains 3D shapes (meshes) captured using multi-camera systems. Topological differences arise when the limbs of the dancer touch each other or the body during motion. We consider three different shape pairs from this sequence in order to illustrate the limitation of our approach.

We follow the same experimental steps in Genctav et al. [6]: We visually compare matching results and quantify matching error or goodness via several measures. The matching error denoted by $\widetilde{D}_{\text{grd}}$ is a normalized deviation from the ground truth correspondence. The lower the $\widetilde{D}_{\text{grd}}$, the better the match. Given a mapping f obtained between sample points of two shapes, the deviation from the ground truth correspondence g is computed as average of the geodesic distances between the points $f(s_i)$ and $g(s_i)$ on the second shape where s_i denotes ith sample point of the first shape. The normalization is performed by dividing the deviation to the sampling radius. $\widetilde{D}_{\text{grd}} \leq 1$ holds for the optimal mapping since it means that the deviation from the ground truth correspondence is within the sampling radius. We report several statistics on $\widetilde{D}_{\text{grd}}$. Also, as a discrete measure of goodness, we count the number of matching results for which $\widetilde{D}_{\text{grd}}$ is less than 1. The larger this count, the more successful is the matching approach. The matching results are also compared with biharmonic-based mapping in which shapes are matched without deflation using biharmonic distance [8]. Note that biharmonic distance is a diffusion-based distance measure which is robust to small topological noise.

6.4 Results and Discussion

6.4.1 Samba Sequence

For the pair in Figure 6.1, the matching result is visually presented in Figure 6.7 (a). The input meshes represent the visual hull of the two shapes obtained by a simple voxel carving algorithm. The models have different topologies as the arms of the second shape touch the body. Our approach utilizes topologically similar representations of the input shapes and provides a mapping between the deflated

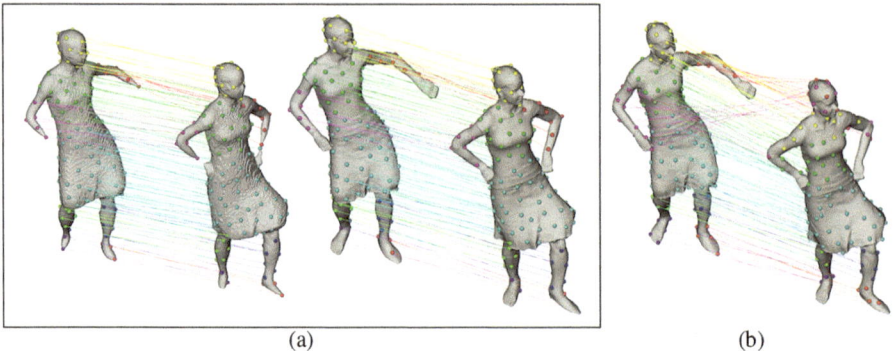

Fig. 6.7 (**a**) Mapping between iso-surfaces at $\ell = 0.052$ (first) and its transfer to the input shapes (second). (**b**) Biharmonic-based mapping

Table 6.2 Performance of the proposed approach in comparison with the biharmonic-based mapping using SHREC10 correspondence benchmark. The results represent average of $\widetilde{D}_{grd}$ over the mappings. The highest topology noise strength is different for the pairs considered at each row

Max noise strength	Biharmonic	Proposed
1	1.67	**1.11**
2	1.69	**1.12**
3	1.70	**1.12**
4	2.22	**1.12**
5	2.53	**1.12**

forms of the input models which are the iso-surfaces at $\ell = 0.052$. The topological similarity between the deflations enables the correct mapping for which $\widetilde{D}_{grd}$ is 0.85. For comparison, the biharmonic distance-based mapping without deflation is given in Figure 6.7 (b) where the right arm of the first model is matched to the head of the second one increasing $\widetilde{D}_{grd}$ to 2.82.

6.4.2   SHREC10

We search a match between each base shape and each of its five deformations containing topological noise. The topological noise is caused by the edge links between different parts of the shapes. We examine the performance of the proposed approach in comparison with the biharmonic-based mapping while the noise strength increases. In Table 6.2, we present average of $\widetilde{D}_{grd}$ over the obtained results where the highest topology noise strength is different for the pairs considered at each row. The biharmonic-based mapping diverges from being optimal when the noise strength is greater than three. Our proposed approach is robust to the topological noise as all of the mappings are very close to the optimal and it performs the best for

Fig. 6.8 Mapping obtained
by our method for two shapes
from SHREC10
correspondence benchmark
where the model of the sitting
man has topological noise of
degree five

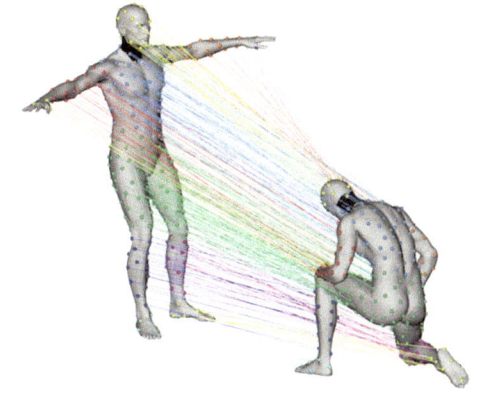

all of the experiments. In Figure 6.8, we show the matching result obtained by our method for two human shapes where the model of the sitting man has topological noise of degree five caused by the edge links between the hands and the legs.

6.4.3 SHREC11

We use the shape models 0002, 0004, 0005, 0007, 0008, 0012, and 0014 for which the ground truth correspondence is available. We use the isometric shape and five shapes with topology noise from each model. We also use the base shape from the models 0002 and 0007. In Figure 6.9, we present $\widetilde{D}_{grd}$ for the mappings obtained using the proposed approach and the biharmonic-based mapping. The input pairs are the shapes from each model where at least one of them has topological distortion. Note that we do not present $\widetilde{D}_{grd}$ for some of the mappings where the symmetric flip problem arises. The number of pairs with the symmetric flip problem is 3 for the proposed method and 12 for the biharmonic mapping. As shown in Figure 6.9 and summarized in Table 6.3, our approach successfully handles the topology noise as almost all of our mappings are optimal ($\widetilde{D}_{grd} \leq 1$). Excluding the mappings with symmetric flip, average of $\widetilde{D}_{grd}$ over all results, avg($\widetilde{D}_{grd}$), is very small for our method compared to the biharmonic-based mapping (see Table 6.3). In Figure 6.10, we present the mappings obtained by the proposed approach for three pairs of shapes.

Finally, we present a visual comparison of our approach with the method [9] which performed the best in the topology noise category of SHREC10 correspondence benchmark. We run the method [9] on a pair of horse shapes from SHREC11 robustness benchmark using its code available on the web. One of the horse shapes has the topological noise as its back legs are linked to each other. Figure 6.11 shows that our approach successfully handles the topology noise, whereas [9] fails to solve the correspondence problem under the given topology noise.

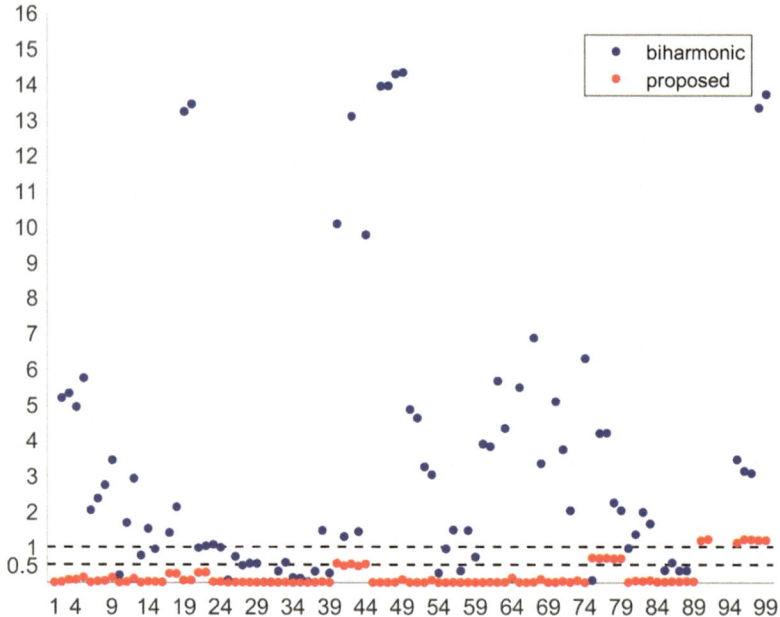

Fig. 6.9 Normalized average ground truth error $\widetilde{D}_{grd}$ for the mappings between topologically different pairs of shapes from SHREC11 robustness benchmark. The errors in blue color are obtained using biharmonic-based mapping. The errors in red color show the performance of the proposed approach. $\widetilde{D}_{grd}$ is not presented for some of the mappings where the symmetric flip problem arises

Table 6.3 Summary of the results in Figure 6.9. Performance of the proposed approach in comparison with the biharmonic-based mapping using SHREC11 robustness benchmark

	Biharmonic	Proposed
# of $\widetilde{D}_{grd} \leq 1$	31	89
avg($\widetilde{D}_{grd}$)	3.45	0.18
stddev($\widetilde{D}_{grd}$)	4.08	0.34
min($\widetilde{D}_{grd}$)	0.01	0
max($\widetilde{D}_{grd}$)	14.32	1.21

6.4.4 Flashkick Sequence

In Figure 6.12, we present the mappings obtained using our approach between three different pairs of shapes from the flashkick sequence [13]. For the pair on the top left, our approach performs well since the topological noise is caused by localized touches, i.e., foot-to-foot and hand-to-leg connection in the first and the second shape, respectively. The pair on the top right represents an input case where our approach starts to fail where the legs in the first shape are merged along their bottom half. The degradation of the matching result, which is especially around the legs, is

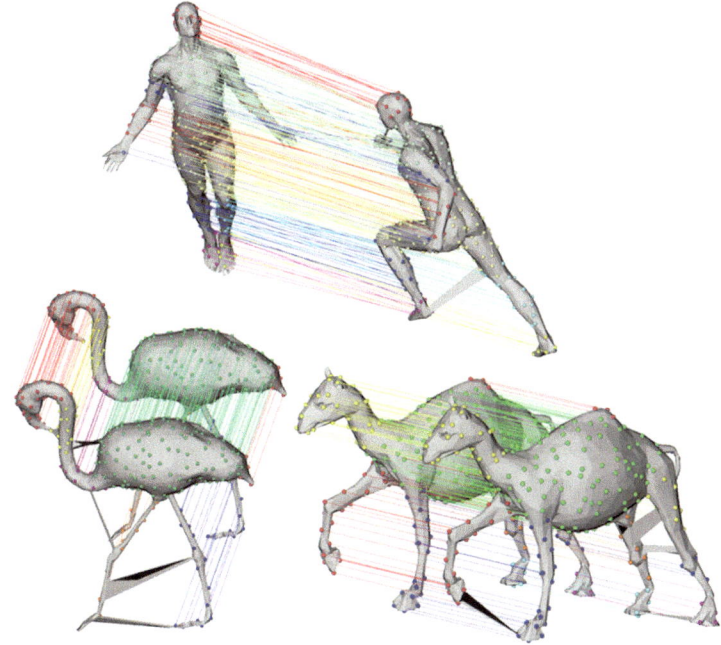

Fig. 6.10 Mappings obtained by the proposed approach for three pairs of shapes from SHREC11 robustness benchmark

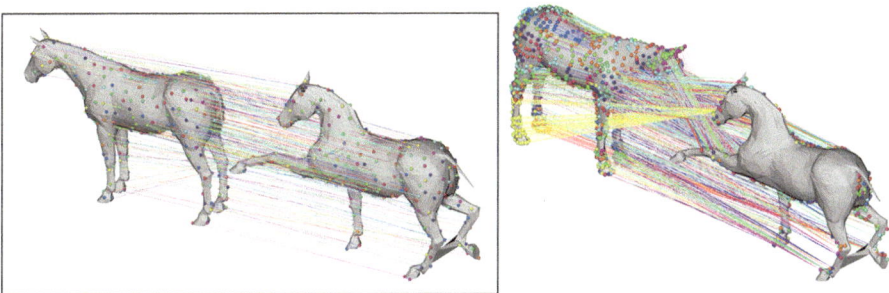

Fig. 6.11 For a pair of shapes from SHREC11 robustness benchmark. (Left) Mapping result obtained by the proposed approach. (Right) Dense mapping result obtained by the method [9]

due to that the topological difference is resolved when the thinner leg is split by the deflations. For the pair at the bottom, our approach results in an incorrect mapping since the corresponding topological difference, which is due to full merge of the legs in the first shape, cannot be removed via deflating the shape boundary.

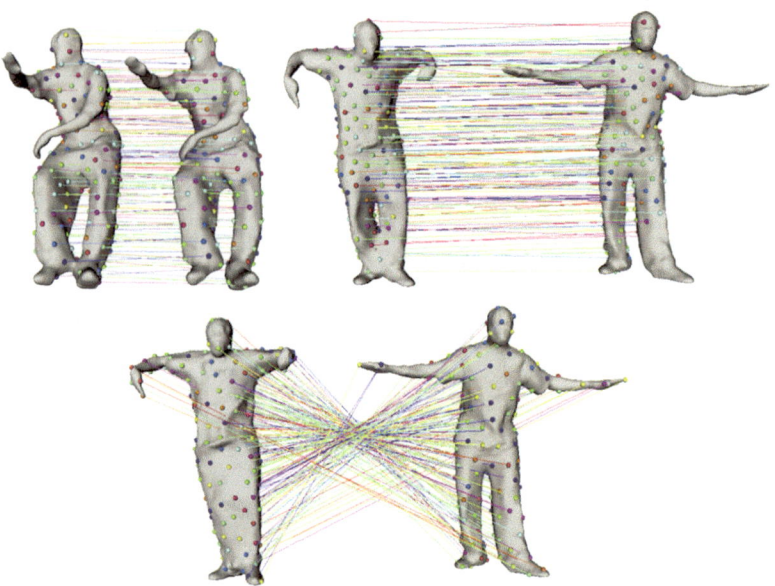

Fig. 6.12 Mappings obtained by the proposed approach for three different pairs of shapes from the flashkick sequence [13]

6.5 Summary and Conclusion

We presented a strategy to bring two shapes to be matched to a comparable topology prior to correspondence search for the purpose of coping with topological noise. Deflations are automatically stopped at an appropriate level by a barrier structure computed for each shape. The topological noise we dealt with includes localized touch of articulations on the other shape regions or spurious links between different shape parts. The robustness of our method is demonstrated with the illustrative experiments using different datasets.

Compared to Genctav et al. [6], one of the immediate advantages of the presented one is the automatic stopping condition used for equating topologies of the input shapes via the barrier structure. Another advantage is a simplified selection of discrete adaptive deflation levels by simply dividing the range of the ρ^0 into equal increments. We showed that the topology equating process takes less number of steps when ρ^0 is used for extracting shape deflations.

Acknowledgements The work is funded by TUBITAK under grant 112E208. We thank Adrian Hilton for sharing the flashkick sequence [13] and i3DPost Multi-View Human Action datasets [7].

References

1. Ahmed, N., Theobalt, C., Rossl, C., Thrun, S., Seidel, H.P.: Dense correspondence finding for parametrization-free animation reconstruction from video. In: Proceedings of Computer Vision and Pattern Recognition (CVPR) (2008)
2. Bronstein, A., Bronstein, M., Kimmel, R., Mahmoudi, M., Sapiro, G.: A Gromov-Hausdorff framework with diffusion geometry for topologically-robust non-rigid shape matching. Int. J. Comput. Vis. **89**(2), 266–286 (2010)
3. Bronstein, A., Bronstein, M., Castellani, U., Falcidieno, B., Fusiello, A., Godil, A., Guibas, L., Kokkinos, I., Lian, Z., Ovsjanikov, M., Patane, G., Spagnuolo, M., Toldo, R.: SHREC 2010: robust large-scale shape retrieval benchmark. In: Proceedings of Eurographics Workshop on 3D Object Retrieval (2010)
4. Bronstein, A., Bronstein, M., Castellani, U., Dubrovina, A., Guibas, L., Horaud, R., Kimmel, R., Knossow, D., Von Lavante, E., Mateus, D., Ovsjanikov, M., Sharma, A.: SHREC 2010: robust correspondence benchmark. In: Proceedings of Eurographics Workshop on 3D Object Retrieval (2010)
5. Genctav, M., Genctav, A., Tari, S.: NonLocal via local–nonlinear via linear: a new part-coding distance field via screened Poisson Equation. J. Math. Imaging Vis. **55**(2), 242–252 (2016)
6. Genctav, A., Sahillioglu, Y., Tari, S.: 3D shape correspondence under topological noise. In: 24th Signal Processing and Communication Application Conference (SIU), pp. 401–404 (2016) (Preprint in English arXiv:1705.00274)
7. Gkalelis, N., Kim, H., Hilton, A., Nikolaidis, N., Pitas, I.: The i3DPost multi-view and 3D human action/interaction. In: Proceedings of CVMP, pp. 159–168 (2009)
8. Lipman, Y., Rustamov, R., Funkhouser, T.: Biharmonic distance. ACM Trans. Graph. **29**(3), 27:1–27:11 (2010)
9. Mateus, D., Horaud, R., Knossow, D., Cuzzolin, F., Boyer, E.: Articulated shape matching using Laplacian eigenfunctions and unsupervised point registration. In: Proceedings of Computer Vision and Pattern Recognition (CVPR) (2008)
10. Ovsjanikov, M., Merigot, Q., Memoli, F., Guibas, L.: One point isometric matching with the heat kernel. Comput. Graph. Forum **29**(5), 1555–1564 (2010)
11. Sharma, A., Horaud, R.: Shape matching based on diffusion embedding and on mutual isometric consistency. In: Proceedings of Computer Vision and Pattern Recognition Workshops (2010)
12. Sharma, A., Horaud, R., Cech, J., Boyer, E.: Topologically-robust 3D shape matching based on diffusion geometry and seed growing. In: Proceedings of Computer Vision and Pattern Recognition (CVPR), pp. 2481–2488 (2011)
13. Starck, J., Hilton, A.: Surface capture for performance based animation. IEEE Comput. Graph. Appl. **27**(3), 21–31 (2007)
14. Tari, S.: Hierarchical shape decomposition via level sets. In: ISMM, pp. 215–225. Springer, Berlin (2009)
15. Tari, S.: Extracting parts of 2D shapes using local and global interactions simultaneously. In: Chen, C.H. (ed.) Handbook of Pattern Recognition and Computer Vision, 4th edn., pp. 283–303. World Scientific, Hackensack, NJ (2009)
16. Tari, S., Shah, J., Pien, H.: A computationally efficient shape analysis via level sets. In: IEEE Workshop on Mathematical Methods in Biomedical Image Analysis (1996)
17. Vlasic, D., Baran, I., Matusik, W., Popovic, J.: Articulated mesh animation from multi-view silhouettes. ACM Trans. Graph. **27**(3), 97:1–97:9 (2008)

Chapter 7
Joint Segmentation and Nonlinear Registration Using Fast Fourier Transform and Total Variation

Thomas Atta-Fosu and Weihong Guo

Abstract Image segmentation and registration play active roles in machine vision and medical image analysis of historical data. We explore the joint problem of segmenting and registering a template (current) image given a reference (past) image. We formulate the joint problem as a minimization of a functional that integrates two well-studied approaches in segmentation and registration: geodesic active contours and nonlinear elastic registration. The template image is modeled as a hyperelastic material (St. Venant-Kirchhoff model) which undergoes deformations under applied forces. To segment the deforming template, a two-phase level set-based energy is introduced together with a weighted total variation term that depends on gradient features of the deforming template. This particular choice allows for fast solution using the dual formulation of the total variation. This allows the segmenting front to accurately track spontaneous changes in the shape of objects embedded in the template image as it deforms. To solve the underlying registration problem, we use gradient descent and adopt an implicit-explicit method and use the fast Fourier transform.

7.1 Introduction

Image registration and segmentation have become fundamental to medical image processing. Respective treatments of these problems have also seen significant contributions, and several models have been proposed to combine the two objectives and take advantage of the respective methods. Since the early works of Yezzi [17], variational models for joint segmentation and registration methods have gained popularity in the medical image processing community. A commonly used approach in the variational framework is to formulate the joint problem using a joint energy

T. Atta-Fosu · W. Guo (✉)
Department of Mathematics, Applied Mathematics & Statistics, Case Western Reserve University, Cleveland, OH, USA
e-mail: thomas.atta-fosu@case.edu; wxg49@case.edu

© The Author(s) and the Association for Women in Mathematics 2018
A. Genctav et al. (eds.), *Research in Shape Analysis*, Association for Women in Mathematics Series 12, https://doi.org/10.1007/978-3-319-77066-6_7

composed of segmentation and registration terms, which are usually interdependent through the unknowns.

The particular connection between the segmentation and registration terms and their contributions to the joint energy characteristically depend on the nature of the problem one wishes to solve. For instance, if a segmentation of the reference (fixed) image is available and the objective is to register and segment a given template image, the mask (or contours) of the segmented regions in the reference image may be used as a data term to fit a segmenting contour of the template image and add a penalty term that encodes the desired registration. An example of this is the work of Vese and Guyader in [12] which combines an active contour-based data term and a nonlinear elasticity-based penalty term into the joint energy (this particular model will be reviewed shortly).

In the case that a prior segmentation of the reference is unavailable, it is not uncommon to use a full registration functional together with a segmentation term in the joint energy (for instance, see [13] although the authors segment the reference image while registering the template). In this case a data term for the registration such as *sum of squared difference*, *mutual information* [7, 16], or *normalized gradients* [10, 14] is combined with a regularizer such as the stored energy density function of a St. Venant-Kirchhoff material (reviewed below) to form the full registration term. This registration term is then complimented by a segmentation term which encodes the desired segmentation of the template.

The ideas presented in this work deal with the joint segmentation and registration framework that combines a nonlinear elasticity-based smoother and the weighted total variation (TV)-based active contour model by Bresson et al. [3]. The proposed method aims to simultaneously register and segment a given template image by taking advantage of the fast TV-based segmentation algorithm as well as an efficient treatment of the nonlinear elasticity-based regularizer (St. Venant-Kirchhoff material model) and its related descent equation. We directly derive the Euler-Lagrange equations in the registration subproblem and decompose it into linear and nonlinear components to utilize a semi-implicit method for solving for the displacement. Imposing periodic boundary conditions on the displacements, we take advantage of the inherent structure present in the discretized linear component of the Euler-Lagrange equation and apply the fast Fourier transform to compute efficient solution of the system. To the best of our knowledge, this is the first time such approach has been used in the joint registration and segmentation framework.

The weighted TV penalty was adopted in the segmentation component due to its inherent numerically fast-converging solution associated with the method of solution as presented in [3]. It also allows the segmenting front to track objects in the deforming template by tracking sudden gradient changes in the template as it deforms. These gradient changes are detected by an edge-stopping function appearing as weight in the weighted TV term. The nonlinear elasticity-based regularizer also ensures that large deformations are admissible in the deforming template. This is in contrast with the linear elasticity-based regularizer which assumes a strain energy density determined by infinitesimal displacement in the template image. This contrast follows from the different strain tensors used in the stored energy density. For a linear elastic material model, the strain tensor is the engineering strain, which is defined as

$$\varepsilon_{\mathbf{u}} = \frac{1}{2} \left(\mathbf{u} + \mathbf{u}^T \right), \tag{7.1}$$

while the strain tensor in the St. Venant-Kirchhoff material model is defined as

$$\mathbf{E_u} = \frac{1}{2} \left(\nabla \mathbf{u} + \nabla \mathbf{u}^T + \nabla \mathbf{u}^T \nabla \mathbf{u} \right), \tag{7.2}$$

where $\mathbf{u}$ is the displacement vector which induces a spatial correspondence between the template and reference images.

In the next subsection, we review some related work on the joint segmentation and registration problem.

7.1.1 Related Works on Joint Segmentation and Registration

Since the work of Yezzi et al. in [17], other works have been done in the joint segmentation and registration framework. In their pioneering work, Yezzi et al. combined an *active contour* segmentation model with a rigid registration framework that sought to simultaneously segment both MRI and CT images. In this framework, the segmenting contour $\hat{C}$ of one image $\hat{I}$ is related to the segmenting curve C of image I through a rigid transformation g:

$$\hat{C} := g(C). \tag{7.3}$$

The objective of the model was to have the two curves segment similar objects in $\hat{I}$ and I. The rigid transform g was taken to be a composition of a rotation $\mathcal{R}_\theta$ and translation $\mathcal{T}$ of local coordinates $\mathbf{x}$ in I:

$$g(\mathbf{x}) = \mathcal{R}_\theta(\mathbf{x}) + \mathcal{T}, \quad \mathcal{R}_\theta = \begin{bmatrix} \cos\theta & \sin\theta \\ -\sin\theta & \cos\theta \end{bmatrix}, \quad \mathcal{T} = \begin{bmatrix} t_x \\ t_y \end{bmatrix}. \tag{7.4}$$

While this framework is elegant in novelty, it was only limited to rigid body displacements, and new models have since been proposed to tackle nonrigid registration and segmentation.

Later in [12], Guyader and Vese proposed a joint registration and segmentation framework with the following problem statement: Given a segmentation of the template image in terms of level set function Φ_0, find a segmentation of the reference image $R(\mathbf{x})$ while registering the template image. The segmentation model adopted by the authors is the piecewise constant ACWE model [5]. The solution $\mathbf{u}$ in the registration component is regularized using the nonlinear elasticity regularizer. In developing their model, the authors introduce an *auxiliary* variable V, which approximates the gradient of the displacement field $\mathbf{u}$. This allowed the authors to circumvent solving the second-order nonlinear Euler-Lagrange equations typically associated with the nonlinear elasticity regularizer.

The proposed energy in their model is given by

$$\mathcal{E}(c_1, c_2, \mathbf{u}, V) = \nu_1 \int_\Omega |R(\mathbf{x}) - c_1|^2 H(\Phi_0(\mathbf{x} - \mathbf{u}(\mathbf{x}))) \, d\mathbf{x}$$

$$+ \nu_2 \int_\Omega |R(\mathbf{x}) - c_2|^2 (1 - H(\Phi_0(\mathbf{x} - \mathbf{u}(\mathbf{x})))) \, d\mathbf{x}$$

$$+ \int_\Omega W(\widehat{V}) \, d\mathbf{x} + \frac{\alpha}{2} \int_\Omega \| \nabla \mathbf{u} - V \|_F^2 \, d\mathbf{x}, \qquad (7.5)$$

where

$$W(\widehat{V}) = \frac{\lambda}{2} tr(\widehat{V})^2 + \mu tr(\widehat{V}^2), \quad \widehat{V} = \frac{V^T + V + V^T V}{2},$$

λ, μ are lamé parameters, and H is the Heaviside function.

The authors used sequential minimization in $\mathbf{u}$, V_{11}, V_{12}, V_{21}, V_{22}, c_1, and c_2 by driving the respective Euler-Lagrange equations to stationary solutions. The initial curve (a level set of Φ_0) which segments objects in the template is driven by the deformation until it aligns with the corresponding objects in the reference image. A major drawback to this method is the computational bottlenecks owing to the introduction of the matrix variable V which has d^2 components each of similar size as the input d−dimensional image.

The recent work by Ozeré et al. in [13] presents a method of segmenting the reference image while registering the template image. The authors use a regularization term which is the sum of the nonlinear elasticity smoother and a penalty term that ensures the Jacobian determinant remains close to 1:

$$S[\mathbf{u}] = \int_\Omega \underbrace{W(\mathbf{E_u}) + \mu (\det F - 1)^2}_{\widehat{W}} d\mathbf{x}, \ F = I + \nabla\mathbf{u}. \qquad (7.6)$$

Existence of global minimizers for functionals involving regularization in (7.6) is not guaranteed, since the strain energy density function of the *St. Venant-Kirchhoff* material is known to be not quasi-convex [15]. The authors therefore replace $\widehat{W}$ with its quasi-convex envelope $Q\widehat{W}$ and decouple the deformation ψ from its gradient by introducing auxiliary variable V (and $\widetilde{T}$) to approximate $\nabla\psi(= F)$ (and $T \circ \psi$, respectively). The authors use a weighted TV penalty, with weight function $g(x) = \frac{1}{1+\beta x^2}$, which attracts level sets in $\widetilde{T}$ to edges in R. The proposed functional to minimize is then given as

$$\mathcal{E}(\psi, V, \widetilde{T}) = \int_\Omega g(|R|)|\nabla\widetilde{T}| + \frac{\nu}{2} \| T \circ \psi - R \|_{L^2(\Omega)}^2 + \int_\Omega Q\widehat{W}(V) d\mathbf{x}$$

$$+ \frac{\gamma}{2} \| V - \nabla\psi \|_{L^2(\Omega, M_2)}^2 + \gamma \| \widetilde{T} - T \circ \psi \|_{L^1(\Omega)}.$$

In the above functional, the authors also use the dual formulation of the weighted *TV* as done by Bresson et al. in [3]. This dual formulation of the *TV* is also adopted in this work.

One caveat of the resulting numerical minimization scheme of the above model comes from the introduction of the auxiliary variable V; As pointed out earlier, there are d^2 components in V, where d is the image dimensions, and one has to solve for all such components in addition to the original variables $\mathbf{u}$ and ψ.

In contrast to the works discussed above, our focus is on segmenting and registering the template image without a reference segmentation.[1] A perceived difficulty associated with the simultaneous segmentation and registration of a template image is that during iterations the template undergoes deformations, and the segmenting contour must adapt to the constantly changing topology of embedded objects in the template.

In addition, we also derive directly the Euler-Lagrange equations of the nonlinear elasticity-based registration model and decompose into linear and nonlinear differential operators. This allows us to apply an implicit-explicit finite difference method to the system, which we solve using the fast Fourier transform.

The remainder of the paper is organized as follows: In Section 7.2 we introduce the mathematical model of proposed framework. Mathematical derivation of relevant equations and resulting numerical schemes are presented in Section 7.3. Numerical experiments are shown in Section 7.4.

7.2 Proposed Framework

In this section we present the proposed simultaneous segmentation and registration method. The model considered is the minimization of a joint energy comprising of two terms, the segmentation and registration energies:

$$\min_{\mathbf{u},\phi}\left\{ \mathcal{E}\left(\mathbf{u},\phi\right) = \mathcal{J}_{reg}\left(T,R,\mathbf{u}\right) + \mathcal{J}_{seg}\left(T,\phi,\mathbf{u}\right)\right\}, \tag{7.7}$$

where $\mathbf{u}$ is the displacement vector which registers the template image T to the reference image R and ϕ is the level set function that segments T. $\mathcal{J}_{reg}$ is the functional for the registration model, and $\mathcal{J}_{seg}$ is the functional for the segmentation model for T. Here we observe that the segmentation functional is dependent on both the segmenting function ϕ and the displacement field $\mathbf{u}$. In what follows we define the two components of $\mathcal{E}$.

---

[1]Useful target application typically involves atlas-based segmentation, and we see this work as a proof of concept to be built upon in the future.

7.2.1 Registration Model

For simplicity we assume that the template and reference images are acquired from the same imaging modality, and we therefore use the *sum of squared difference* dissimilarity measure as the data term in $\mathcal{J}_{reg}$:

$$Dist\,(T\,(\mathbf{x} + \mathbf{u}),\,R(\mathbf{x})) = \frac{1}{2} \int_{\Omega} (T\,(\mathbf{x} + \mathbf{u}) - R(\mathbf{x}))^2\,d\mathbf{x}. \tag{7.8}$$

Moreover, in order that large deformation of the template image is admissible, we adopt the stored strain energy density, W, of the *St. Venant-Kirchhoff* material:

$$W\,(\mathbf{E_u}) = \frac{\lambda}{2} \left(tr\,(\mathbf{E_u}) \right)^2 + \mu tr\left(\mathbf{E_u^2}\right), \tag{7.9}$$

Thus the component $\mathcal{J}_{reg}$ of (7.7) is given as

$$\mathcal{J}_{reg}\,(T, R, \mathbf{u}) = \frac{1}{2} \int_{\Omega} (T_{\mathbf{u}} - R(\mathbf{x}))^2\,d\mathbf{x} + \int_{\Omega} W\,(\mathbf{E_u})\,d\mathbf{x}, \tag{7.10}$$

where

$$T_{\mathbf{u}} = T\,(\mathbf{x} + \mathbf{u})\,.$$

7.2.2 Segmentation Model

For the segmentation term $\mathcal{J}_{seg}$, we introduce the weighted total variation penalty model of Bresson et al. [3]. In [3], the authors introduced a convex-relaxed form of the Chan-Vese model [5], and the model is given by

$$\min_{0 \leq \phi \leq 1} \left\{ F\,(c_1, c_2, \phi) = TV_g(\phi) + \xi \int_{\Omega} r(x, c_1, c_2)\phi(x)dx \right\}, \tag{7.11}$$

where

$$r(x, c_1, c_2) = (u_0 - c_1)^2 - (u_0 - c_2)^2\,,$$

u_0 is the image to segment, and ϕ is a level set function whose η-level set segments u_0. The unknowns c_1 and c_2 are the respective mean values of u_0 inside and outside the region demarcated by the η-level set of ϕ, where $\eta \in (0, 1)$. The parameter ξ is a positive number that controls the smoothness or tightness of the segmenting contour, with smaller ξ leading to smoother segmenting contours. The weighted total variation term $TV_g(\phi)$ is defined as

$$TV_g(\phi) = \int_\Omega g(u_0)|\nabla\phi|dx, \tag{7.12}$$

and g is the edge-stopping function

$$g(u_0) = \frac{1}{1 + \beta|\nabla u_0|^2},$$

where $\beta > 0$ accounts for significance of the edge weight – the higher the value of β, the lesser the weight of edges in stopping ϕ near edges in the image u_0.

To integrate the above model into the joint energy, we define the segmentation term as follows:

$$\mathcal{J}_{seg}(T, \phi, c_1, c_2, \xi) = \int_\Omega g(T_\mathbf{u})|\nabla\phi|dx + \xi \int_\Omega r(\mathbf{u}, c_1, c_2)\phi dx, \tag{7.13}$$

where

$$r(\mathbf{u}, c_1, c_2) = \left((c_1 - T_\mathbf{u})^2 - (c_2 - T_\mathbf{u})^2\right), \quad \text{and } 0 \le \phi \le 1.$$

Observe that in (7.13), we have absorbed the coordinate variable x, which appeared in r (7.11), into the displacement vector $\mathbf{u}$ which depends on x.

7.3   Mathematical Derivation, Implementation, and Numerical Experiments

We use alternating minimization [8] to obtain optimal displacements $\mathbf{u}$ and segmenting front ϕ which minimize (7.7). We first minimize $\mathcal{E}(\mathbf{u}, \phi)$ with respect to $\mathbf{u}$, and then with respect to ϕ, and iterate until convergence. The order of minimizing first in $\mathbf{u}$ and then in ϕ has a practical consequence. Microstructures in the template image change in response to the new displacement values, and the segmenting front is also impacted by sudden topological changes. The sequence of updates is therefore carried out this way so that the segmenting front ϕ is able to instantly track changes occurring in the deforming template after the template has been updated according to new displacement values.

7.3.1 Optimal Displacement Subproblem

Since the two terms $\mathcal{J}_{reg}$ and $\mathcal{J}_{seg}$ both depend on the displacement $\mathbf{u}$, the displacement subproblem remains as the minimization of $\mathcal{E}$ with respect to $\mathbf{u}$:

$$\min_{\mathbf{u}} \frac{1}{2} \int_{\Omega} (T_{\mathbf{u}} - R(\mathbf{x}))^2 \, d\mathbf{x} + \int_{\Omega} W(\mathbf{E}_{\mathbf{u}}) \, d\mathbf{x} + T V_g(\phi) + \xi \int_{\Omega} r(\mathbf{u}, c_1, c_2) \phi d\mathbf{x}. \tag{7.14}$$

From elementary calculus of variation, we have the following Euler-Lagrange equation:

$$-\nabla \cdot \left((I + \nabla \mathbf{u}) \, S \right) + \nabla T_{\mathbf{u}} \left(T_{\mathbf{u}} - R(\mathbf{x}) \right) -$$
$$\partial_{\mathbf{u}} \tilde{g} - 2\xi \nabla T_{\mathbf{u}} \left[(c_1 - T_{\mathbf{u}}) - (c_2 - T_{\mathbf{u}}) \right] \phi = 0 \tag{7.15}$$

where

$$S = \frac{\partial W}{\partial \mathbf{E}_{\mathbf{u}}} = \lambda \mathrm{tr} \, (\mathbf{E}_u) \, I + 2\mu \mathbf{E}_{\mathbf{u}}. \tag{7.16}$$

$$\partial_{\mathbf{u}} \tilde{g} = 2\beta \left[\frac{|\nabla \phi|}{\left(1 + \beta |\nabla T_{\mathbf{u}}|^2 \right)^2} \right] \nabla^2 T_{\mathbf{u}} \nabla T_{\mathbf{u}}.$$

A close examination of the first term in (7.15) reveals that it is a composite of linear and nonlinear second-order differential operators. We exploit this "stroke-of-luck" and write it in terms of $\mathcal{L}$ and $\mathcal{N}$, respectively, for the linear and nonlinear operators:

$$\nabla \cdot \left((I + \nabla \mathbf{u}) \, S \right) = \mathcal{L}[\mathbf{u}] + \mathcal{N}[\mathbf{u}], \tag{7.17}$$

where

$$\mathcal{L}[\mathbf{u}] = \mu \Delta \mathbf{u} + (\lambda + \mu) \nabla (\nabla \cdot \mathbf{u}), \tag{7.18}$$

$$\mathcal{N}[\mathbf{u}] = \nabla \cdot ((\nabla \mathbf{u}) \, S) + \nabla \cdot \left[\frac{\lambda}{2} \mathrm{tr} \left((\nabla \mathbf{u})^T \nabla \mathbf{u} \right) I + \mu \left((\nabla \mathbf{u})^T \nabla \mathbf{u} \right) \right]. \tag{7.19}$$

Hence (7.15) can be written as

$$\mathcal{L}[\mathbf{u}] + \mathcal{N}[\mathbf{u}] - \mathbf{f}(\mathbf{u}, T_{\mathbf{u}}, c_1, c_2, \phi) = 0. \tag{7.20}$$

where

$$\mathbf{f}(\mathbf{u}, T_{\mathbf{u}}, c_1, c_2, \phi) = \nabla T_{\mathbf{u}} (T_{\mathbf{u}} - R(\mathbf{x})) - \partial_{\mathbf{u}} \tilde{g} - 2\xi \nabla T_{\mathbf{u}} ((c_1 - T_{\mathbf{u}}) - (c_2 - T_{\mathbf{u}})) \phi. \tag{7.21}$$

The operator $\mathcal{L}$ can be expressed in more familiar terms as follows:

$$\mathcal{L}[\mathbf{u}] = \begin{bmatrix} (2\mu + \lambda)\partial_{x_1 x_1} + \mu\partial_{x_2 x_2} & (\lambda + \mu)\partial_{x_1 x_2} \\ (\lambda + \mu)\partial_{x_2 x_1} & (2\mu + \lambda)\partial_{x_2 x_2} + \mu\partial_{x_1 x_1} \end{bmatrix} \begin{bmatrix} \mathbf{u}_1 \\ \mathbf{u}_2 \end{bmatrix}. \tag{7.22}$$

To solve (7.20) numerically, we use a gradient descent method and compute $\mathbf{u}$ as the steady-state solution of the time-dependent PDE:

$$\partial_t \mathbf{u} = \mathcal{L}[\mathbf{u}] + \mathcal{N}[\mathbf{u}] - \mathbf{f}(\mathbf{u}, T_{\mathbf{u}}, c_1, c_2, \phi). \tag{7.23}$$

The solution of (7.23) can be computed by using an *implicit-explicit* (IMEX) scheme (see [1]). We first discretize the solution and then use finite difference methods to approximate the spatial derivatives at the grid points. From here on we will let **L** and **N** denote the finite difference approximations of the operators $\mathcal{L}$ and $\mathcal{N}$, respectively, and for simplicity, we use the notation $\mathbf{u}$ for the vectorized solution after discretization. Then (7.23) becomes

$$\mathbf{u}_t = \mathbf{L}[\mathbf{u}] + \mathbf{N}[\mathbf{u}] - \mathbf{f}(\mathbf{u}, T_{\mathbf{u}}, c_1, c_2, \phi), \tag{7.24}$$

and using the first-order *semi-implicit backward differentiation formula* (SBDF), we have the system

$$\frac{\mathbf{u}^{n+1} - \mathbf{u}^n}{\Delta t} = \mathbf{L}[\mathbf{u}^{n+1}] + \mathbf{N}[\mathbf{u}^n] - \mathbf{f}\left(\mathbf{u}^n, T_{\mathbf{u}}, c_1, c_2, \phi\right), \tag{7.25}$$

where Δt is the time step of the scheme and $\mathbf{u}^n$ represents the solution at time n. In the finite difference discretization of the partial derivatives, we use the following central difference *kernels* and impose periodic boundary conditions:

$$d_1 = d_2 = \frac{1}{2}[1, 0, -1] \tag{7.26}$$

for the first-order partials ∂_{x_1}, ∂_{x_2}, and

$$d_{11} = d_{22} = [1, -2, 1] \tag{7.27}$$

for the second-order partials $\partial_{x_1 x_1}$, $\partial_{x_2 x_2}$.

The scheme leads to a highly structured 2×2 block matrix **L**, whose block components $\mathbf{L}^{11}$, $\mathbf{L}^{22}$, and $\mathbf{L}^{12} = \mathbf{L}^{21}$ are *block-circulant-circulant-block* (BCCB).[2] This fact follows from the periodic boundary conditions imposed on $\mathbf{u}$. In Figure 7.1, examples of the structure in **L** are shown for two images of different sizes.

---

[2]BCCB matrix is a block matrix A whose blocks A_{ij} are circulant matrices and the blocks A_{ij} considered as entries of A form a circulant matrix.

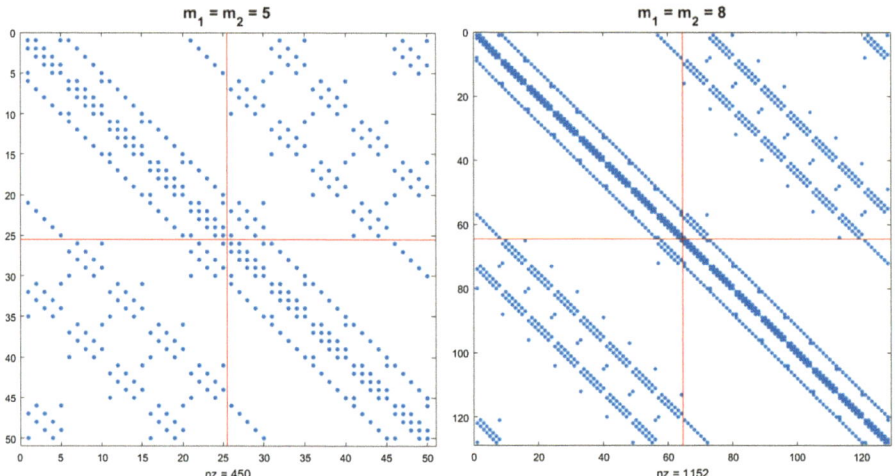

Fig. 7.1 Two structured sparse coefficient matrices of the system (7.22). The panel on the left shows the structure of **L** for a 5 × 5, with 450 nonzero entries out of 2500. In the right panel, the grid size is 8 × 8, with 1150 nonzero entries out of 16384. The percentage of nonzero entries significantly decreases as the number of pixels increases

Since we march forward in time in (7.25), at every iterate n, we solve a system of equations of size $2m_1m_2$ for an image of size $m_1 \times m_2$ ($m_1 \times m_2$ unknowns for each of the two components of **u**). The size of the discretized system (7.25) makes it computationally expensive to solve using iterative methods.

To circumvent this, we take advantage of the structure of **L** and use the fact that each of the four blocks of **L** is **BCCB** and hence is diagonalizable using the 2D Fourier matrix $\mathcal{F}$ and its inverse $\mathcal{F}^*$. See Chapter 4 of [11].

Therefore, we can rewrite **L** as follows:

$$\mathbf{L} = \begin{bmatrix} \mathbf{L}^{11} & \mathbf{L}^{12} \\ \mathbf{L}^{21} & \mathbf{L}^{22} \end{bmatrix} = \begin{bmatrix} \mathcal{F}^* & \mathbf{0} \\ \mathbf{0} & \mathcal{F}^* \end{bmatrix} \begin{bmatrix} \Lambda_{11} & \Lambda_{12} \\ \Lambda_{21} & \Lambda_{22} \end{bmatrix} \begin{bmatrix} \mathcal{F} & \mathbf{0} \\ \mathbf{0} & \mathcal{F} \end{bmatrix}, \tag{7.28}$$

The matrices of eigenvalues Λ_{11}, Λ_{22}, and $\Lambda_{12} = \Lambda_{21}$ of the blocks $\mathbf{L}_{11}$, $\mathbf{L}_{22}$, and $\mathbf{L}_{12}(= \mathbf{L}_{21})$, respectively, can be computed analytically (following the approach in Chapter 5 of [9] using the kernels in (7.26) and (7.27)). These diagonal matrices Λ_{11}, Λ_{12}, and Λ_{22} are reshaped into the same size as the image and are computed as follows:

$$\Lambda_{11}(j,k) = 2(\lambda + 2\mu)(-1 + \cos(2\pi \frac{j}{m_1})) + 2(\lambda + \mu)(-1 + \cos(2\pi \frac{k}{m_2}))$$

$$\Lambda_{22}(j,k) = 2(\lambda + 2\mu)(-1 + \cos(2\pi \frac{k}{m_2})) + 2(\lambda + \mu)(-1 + \cos(2\pi \frac{j}{m_1}))$$

$$\Lambda_{12}(j, k) = \Lambda_{21}(j, k) = -(\lambda + \mu) \sin(2\pi \frac{j}{m_1}) \sin(2\pi \frac{k}{m_2})$$

$$\text{for } j = 0, 1, \ldots, m_1 - 1, \quad k = 0, 1, \ldots, m_2 - 1.$$

Having obtained a representation of $\mathbf{L}$ in (7.28), we now turn our attention to solving the linear system (7.25), which, for brevity, we rewrite as

$$\mathbf{u}^{n+1} - \Delta t \times \mathcal{L}[\mathbf{u}^{n+1}] = \underbrace{\mathbf{u}^n + \Delta t \times \mathbf{N}[\mathbf{u}^n] - \Delta t \times \mathbf{f}\left(\mathbf{u}^n, T_{\mathbf{u}}, c_1, c_2, \phi\right)}_{\mathbf{b}^n}. \quad (7.29)$$

The nonlinear derivatives in $\mathbf{b}^n$ are evaluated through circular convolutions using the appropriate *kernels* defined in (7.26) and (7.27).

The last equation can be expressed compactly in matrix form as

$$\left(\begin{bmatrix} I & 0 \\ 0 & I \end{bmatrix} - \Delta t \begin{bmatrix} \mathbf{L}^{11} & \mathbf{L}^{12} \\ \mathbf{L}^{21} & \mathbf{L}^{22} \end{bmatrix} \right) \begin{bmatrix} \mathbf{u}_1^{n+1} \\ \mathbf{u}_2^{n+1} \end{bmatrix} = \begin{bmatrix} \mathbf{b}_1^n \\ \mathbf{b}_2^n \end{bmatrix}, \quad (7.30)$$

where I is the identity matrix of size $m_1 m_2 \times m_1 m_2$.

Combining (7.28) and (7.30), we have

$$\left(\begin{bmatrix} I & 0 \\ 0 & I \end{bmatrix} - \Delta t \begin{bmatrix} \mathcal{F}^\star & 0 \\ 0 & \mathcal{F}^\star \end{bmatrix} \begin{bmatrix} \Lambda_{11} & \Lambda_{12} \\ \Lambda_{21} & \Lambda_{22} \end{bmatrix} \begin{bmatrix} \mathcal{F} & 0 \\ 0 & \mathcal{F} \end{bmatrix} \right) \begin{bmatrix} \mathbf{u}_1^{n+1} \\ \mathbf{u}_2^{n+1} \end{bmatrix} = \begin{bmatrix} \mathbf{b}_1^n \\ \mathbf{b}_2^n \end{bmatrix}. \quad (7.31)$$

After applying the block Fourier matrix to (7.31), we have the system

$$\begin{bmatrix} 1 - \Delta t \times \Lambda_{11} & -\Delta t \times \Lambda_{12} \\ -\Delta t \times \Lambda_{21} & 1 - \Delta t \times \Lambda_{22} \end{bmatrix} \begin{bmatrix} \widehat{\mathbf{u}}_1^{n+1} \\ \widehat{\mathbf{u}}_2^{n+1} \end{bmatrix} = \begin{bmatrix} \widehat{\mathbf{b}}_1^n \\ \widehat{\mathbf{b}}_2^n \end{bmatrix}, \quad (7.32)$$

where we have defined for the Fourier transform,

$$\widehat{\mathbf{a}} = \mathcal{F}\mathbf{a}.$$

We observe that (7.32) is just a simple 2×2 system with symmetric coefficient and can be written for each pixel (j, k):

$$\underbrace{\begin{bmatrix} 1 - \Delta t \times \Lambda_{11}(j, k) & -\Delta t \times \Lambda_{12}(j, k) \\ -\Delta t \times \Lambda_{21}(j, k) & 1 - \Delta t \times \Lambda_{22}(j, k) \end{bmatrix}}_{\Lambda(i, j)} \underbrace{\begin{bmatrix} \widehat{\mathbf{u}}_1^{n+1}(j, k) \\ \widehat{\mathbf{u}}_2^{n+1}(j, k) \end{bmatrix}}_{\widehat{\mathbf{u}}^{n+1}(j, k)} = \underbrace{\begin{bmatrix} \widehat{\mathbf{b}}_1^n(j, k) \\ \widehat{\mathbf{b}}_2^n(j, k) \end{bmatrix}}_{\widehat{\mathbf{b}}^n(j, k)}. $$

$$(7.33)$$

Before solving (7.33), we observe that the pixels corresponding $j = k = 0$ do not have a unique solution since the reshaped diagonals are all zero whenever $j = k = 0$. With this hindsight, we solve (7.33) for $\widehat{\mathbf{u}}^{n+1}$ using the *Moore-Penrose generalized inverse* of the coefficient $\Lambda(j, k)$. Since Λ is symmetric, we have the following definition for the generalized inverse of Λ:

Definition 7.1 *Let*

$$\Lambda = \begin{bmatrix} a & b \\ b & c \end{bmatrix}.$$

Then the generalized inverse $\Lambda^\dagger$ is given by

$$\Lambda^\dagger = \begin{cases} \dfrac{1}{ac - b^2} \Lambda^{-1} & \text{if } ac - b \neq 0 \\[2mm] \dfrac{1}{a^2 + c^2 + 2b^2} \Lambda^T & \text{if } c = \frac{b^2}{a} \text{ or } a = \frac{b^2}{c} \\[2mm] 0 & \text{otherwise} \end{cases}.$$

The definition for the *generalized inverse* given above is a direct consequence of the Penrose algorithm, which can be found in [9].

After obtaining the *Moore-Penrose inverse* $\Lambda^\dagger$, $\widehat{\mathbf{u}}^{n+1}$ is obtained from (7.33) by the equation

$$\widehat{\mathbf{u}}^{n+1} = \Lambda^\dagger \widehat{\mathbf{b}}^n. \tag{7.34}$$

We finally obtain $\mathbf{u}^{n+1}$ by taking the inverse Fourier transform of $\widehat{\mathbf{u}}^{n+1}$.

7.3.2 Minimization in Segmenting Curve

After obtaining an iterate of $\mathbf{u}$, we solve for an iterate of the segmenting front ϕ and minimize $\mathcal{E}$ with respect to ϕ. In this particular subproblem, we observe that the $\mathcal{J}_{reg}$ is independent of ϕ, and thus the problem reduces to minimizing $\mathcal{J}_{seg}$ in ϕ. For the most part, the content of this section is adapted from [3]. First the constrained minimization is transformed into unconstrained minimization via the following theorem, proof of which can be found in [3].

Theorem 7.1 *Let $r(\mathbf{u}, c_1, c_2) \in L^\infty(\Omega)$, for any given $c_1, c_2 \in \mathbb{R}$, and any scalar $\xi > 0$, the convex constrained minimization problem*

$$\min_{0 \leq \phi \leq 1} \left\{ TV_g(\phi) + \xi \int_\Omega r(\mathbf{u}, c_1, c_2) \phi d\mathbf{x} \right\} \tag{7.35}$$

has the same set of minimizers as the following convex and unconstrained minimization problem:

$$\min_\phi \left\{ TV_g(\phi) + \xi \int_\Omega r(\mathbf{u}, c_1, c_2) \phi d\mathbf{x} + \alpha \int_\Omega \nu(\phi) d\mathbf{x} \right\}, \tag{7.36}$$

where $v(\cdot) := \min\left\{0, 2| \cdot -\frac{1}{2}|\right\}$ is an exact penalty function provided that

$$\alpha > \frac{\lambda}{2} \parallel r(\mathbf{u}, c_1, c_2) \parallel_{L^{\infty}(\Omega)} .$$

As consequence of the last theorem, we solve (7.36) instead of (7.35). To efficiently solve (7.36) using gradient descent, the TV_g term has to be regularized with some parameter $\varepsilon \ll 1$, so that

$$TV_g^{\varepsilon}(\phi) = \int_{\Omega} g(T_{\mathbf{u}})\sqrt{|\nabla\phi|^2 + \varepsilon}\, d\mathbf{x}. \tag{7.37}$$

This is necessary to avoid numerical instabilities when solving for ϕ since the Euler-Lagrange equations for (7.36) is the following PDE:

$$\nabla \cdot \left(g(T_{\mathbf{u}})\frac{\nabla\phi}{|\nabla\phi|}\right) - \xi r(\mathbf{u}, c_1, c_2) - \alpha v'(\phi) = 0, \tag{7.38}$$

and the weighted mean curvature $\nabla \cdot \left(g\frac{\nabla\phi}{|\nabla\phi|}\right)$ is undefined when $|\nabla\phi|$ is zero (when ϕ is a flat surface).

Instead of solving (7.36) with the regularized TV_g, which is sensitive to ε, we solve the following proximal problem (as illustrated in [2–4]):

$$\min_{\phi,\varphi} \left\{TV_g(\phi) + \frac{1}{2\theta} \parallel \phi - \varphi \parallel_{L^2}^2 +\xi \int_{\Omega} r(\mathbf{u}, c_1, c_2)\varphi + \alpha v(\varphi)d\mathbf{x}\right\}, \tag{7.39}$$

where θ controls the proximity of ϕ to φ, so that when θ is small, φ closely approximates ϕ. (See Table 7.1 for list of values of φ as well as other parameters used in the different experiments.)

An alternating minimization approach is used to solve (7.39) iteratively. We first minimize in ϕ, followed by minimizing in φ:

1 Given φ, solve

$$\min_{\phi} \left\{TV_g(\phi) + \frac{1}{2\theta} \parallel \phi - \varphi \parallel_{L^2}^2\right\}, \tag{7.40}$$

Table 7.1 This table shows the parameters used for the different experiments in this section

Data	μ	λ	ξ	θ	β	δt	Δt
Rectangle	500	1	10	2	0.01	0.02	0.01
Cat	500	1	10	20	0.01	0.02	0.01
Cardiac images	5000	1	100	100	0.01	0.02	0.1

2 Given ϕ, solve

$$\min_{\varphi} \left\{ \frac{1}{2\theta} \parallel \phi - \varphi \parallel_{L^2}^2 + \xi \int_{\Omega} r(\mathbf{u}, c_1, c_2)\varphi + \alpha v(\phi) d\mathbf{x} \right\}. \tag{7.41}$$

The solutions to (7.40) and (7.41) are obtained according to the following propositions[3]:

Proposition 7.1 *The solution of (7.40) is given by*

$$\phi = \varphi - \theta \nabla \cdot p, \tag{7.42}$$

where $p = (p_1, p_2)^T$ solves

$$g(T_{\mathbf{u}}) \nabla (\theta \nabla \cdot p - \varphi) - |\nabla (\theta \nabla \cdot p - \varphi)| p = 0. \tag{7.43}$$

The last Equation (7.43) can be solved by a semi-implicit gradient descent method (after introducing an artificial time parametrization of p):

$$p^0 = 0, \quad p^{n+1} = \frac{p^n + \delta t \nabla (\nabla \cdot p^n - \varphi/\theta)}{1 + \frac{\delta t}{g(T_{\mathbf{u}})} |\nabla (\nabla \cdot p^n - \varphi/\theta)|} \; for \, n = 1, 2, \ldots \tag{7.44}$$

where δt is the time step.

Proposition 7.2 *The solution of (7.41) is given by*

$$\varphi = \min \left\{ \max \left\{ \phi - \theta \xi r(\mathbf{u}, c_1, c_2), 0 \right\}, 1 \right\}. \tag{7.45}$$

Based on the iterates p^n in (7.44), the values of ϕ and φ are also updated iteratively according to (7.42) and (7.45).

The values of c_1 and c_2 are updated as the mean values of $T_{\mathbf{u}}$ inside and outside the region determined by the $\eta-$level set, for some $\eta \in (0, 1)$:

$$c_1 = \frac{\int_{\Omega} T_{\mathbf{u}} \phi H(\phi - \eta) d\mathbf{x}}{\int_{\Omega} \phi H(\phi - \eta) d\mathbf{x}}, \quad c_2 = \frac{\int_{\Omega} T_{\mathbf{u}} \phi H(\eta - \phi) d\mathbf{x}}{\int_{\Omega} \phi H(\eta - \phi) d\mathbf{x}}, \tag{7.46}$$

where H is the Heaviside function defined as

$$H(z) = \begin{cases} 1 & \text{if } z \geq 0 \\ 0 & \text{otherwise.} \end{cases}$$

---

[3]Proof of these propositions follows similarly from [3, 4, 6].

7.3.3 Stopping Criteria

In the proposed joint model, the template image T keeps changing, which also affects the shapes of objects to segment within the template until the template is fully registered to the reference. Owing to this, we terminate the iterative process when the registration of T to R is deemed satisfactory.

In our implementation, we use relative changes in the distance between R and $T_{\mathbf{u}}$ as a test for convergence. We define the relative change in the distance between R and $T_{\mathbf{u}}$ at iteration $n + 1$ as

$$\gamma^{n+1} = \text{abs} \left(\frac{\| T_{\mathbf{u}^{n+1}} - R \|_{L^2} - \| T_{\mathbf{u}^n} - R \|_{L^2}}{\| T_{\mathbf{u}^n} - R \|_{L^2}} \right). \tag{7.47}$$

When successive M_γ iterates of γ^n are less than a given tolerance tol, we terminate the iterations. In the experiments presented in the next section, typical value of tol is 1×10^{-3}, and M_γ differs from image to image but ranges between 3 and 7.

To conclude this section, we provide a summary of the algorithm for the joint segmentation and registration model discussed above.

7.4 Results

In this section we provide some experimental results on both academic examples (Figures 7.2 and 7.3) and MRI images of the cardiac area. These experiments were chosen with the view to demonstrating the potential of the algorithm when applied on fundamentally distinct but related images. The second set of experiments were conducted on cardiac images.

In all experiments we show the template and reference images, the deformed template, the segmenting curve (η-level set), and the deformed grid. In the academic examples, iterates of the algorithm are shown, while difference images before and after registration is also shown for the cardiac images.

As explained in [3], during the iterations, the segmenting curve approaches a binary function so that it is almost piecewise constant. Thus the value of η can be selected to be any value in the unit interval. For purposes of consistency, we fix $\eta = 0.5$ for all images.

The next set of figures show the results of the algorithm on images of selected time points during the cardiac cycle (Figures 7.4, 7.5, 7.6, 7.7, 7.8, and 7.9). The template and reference images are two distinct frames (at different time points) of cine cardiac MRI image. In each experiment, the reference image is chosen during ventricular diastole (phase of the cardiac cycle when the ventricle chambers of the heart relaxes), while the template is chosen in the systolic phase (when the chambers contract). A simultaneous registration and segmentation of cardiac images can play important role in cardiovascular applications. During the cardiac cycle, the left ventricle is responsible for pumping blood to the rest of the body. An estimation of

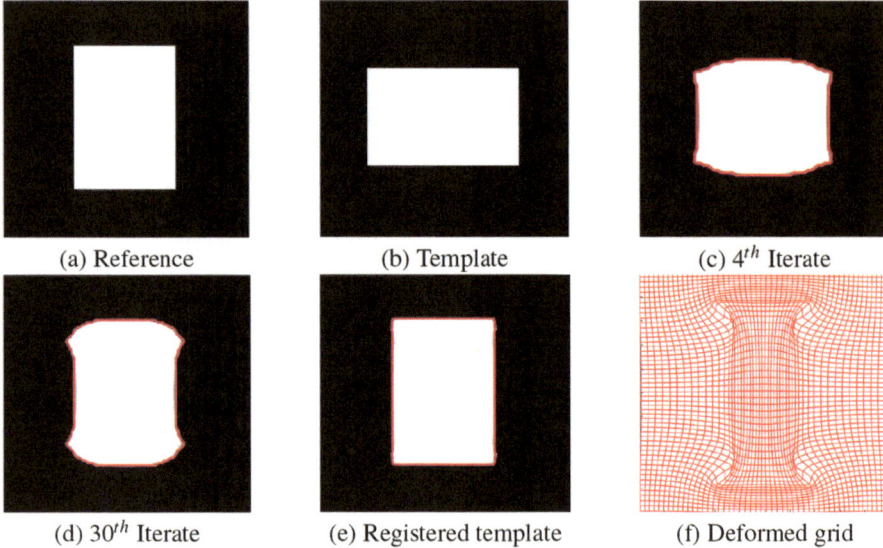

(a) Reference   (b) Template   (c) 4th Iterate

(d) 30th Iterate   (e) Registered template   (f) Deformed grid

Fig. 7.2 The reference and template images are shown in panels (**a**) and (**b**), respectively. The template image is a compressed and stretched version of the reference. In panels (**c**) and (**d**), intermediate results of the deforming template and evolving front are shown for 4th and 30th iterates, respectively. The final segmented and registered template is shown in panel (**e**) with the segmenting level set of ϕ superimposed on the registered image. The deformed grid of T is shown in panel (**f**)

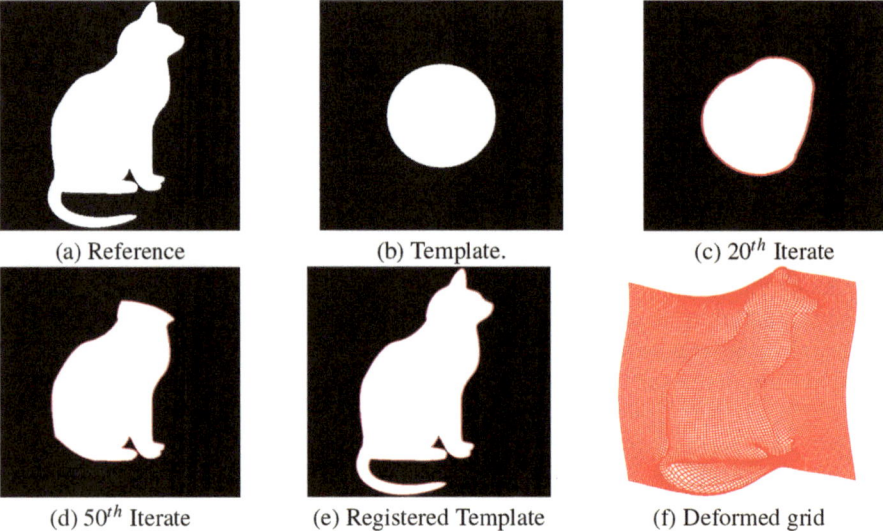

(a) Reference   (b) Template.   (c) 20th Iterate

(d) 50th Iterate   (e) Registered Template   (f) Deformed grid

Fig. 7.3 In this figure we deform a simple 2D disc in panel (**b**) into the shape of a cat in panel (**a**). Intermediate iterates 20 and 30 are shown in panels (**c**) and (**d**), respectively. Although ambitious, the shape of the reference is matched quite well by the deformed template as seen in panel (**e**). The deformed grid in panel (**f**) also shows a substantial change in the grid of the template

---

Algorithm 7.1 Joint segmentation and registration algorithm

Data: T, R, Λ; parameters: λ, μ, ξ, θ, η, Δt, δt, tol, M_γ
Result: Displacement $\mathbf{u}$, Segmentation ϕ, Registered template $T_\mathbf{u}$
   Initialize: $n = 0$, $\mathbf{u}^n = \mathbf{0}$, $\phi^n = 0$, $\varphi^n = 0$, $p^n = \mathbf{0}$, $c_1 = c_2 = 0$, *count* $= 0$
   while count $< M_\gamma$ **do**
      Compute $\mathbf{b}^n$: (7.29)
      $\widehat{\mathbf{b}}^n = FFT2(\mathbf{b}^n)$
      Solve $\Lambda\,\widehat{\mathbf{u}}^{n+1} = \widehat{\mathbf{b}}$ using (7.34)
      Update $\mathbf{u}^{n+1} = \text{IFFT2}(\widehat{\mathbf{u}}^{n+1})$
      Update: Compute $T_{\mathbf{u}^{n+1}}$ by interpolation
      Update: c_1 and c_2 using (7.46)
      Update: φ^{n+1} using (7.45)
      Update: p^{n+1} using (7.44)
      Update: ϕ^{n+1} using (7.42)
      Evaluate γ^{n+1} using (7.47)
      if $\gamma^{n+1} <$ tol **then**
         count $\leftarrow$ count $+1$
      else
         count $\leftarrow 0$
      end if
      if count$\geq M_\gamma$ **then**
         return $\mathbf{u}^{n+1}$, ϕ^n
      else
         Update $n \leftarrow n + 1$
      end if
   end while

---

the total fraction of received blood pumped out of the left ventricle is called *ejection fraction* (EF) and can be a proxy for the health status of the chamber. Registering between diastolic and systolic images yields a deformation field which gives a visual impression of the amount of expansion and contraction of the chamber, while the segmentation of the registered template can help practitioners to zone in on specific structures of interest.

7.5 Conclusion and Discussions

We have presented a fast Fourier transform-based method for joint segmentation and registration method. An important contribution of this work is the direct derivation of Euler-Lagrange equations associated with the registration subproblem. The derivation obviates the need to introduce additional variables as is common in methods using nonlinear elasticity-based regularizers. The direct derivation of the associated Euler-Lagrange equations also allows us to decompose the Euler-Lagrange into linear and nonlinear components, which allows for efficient solution using the fast Fourier transform.

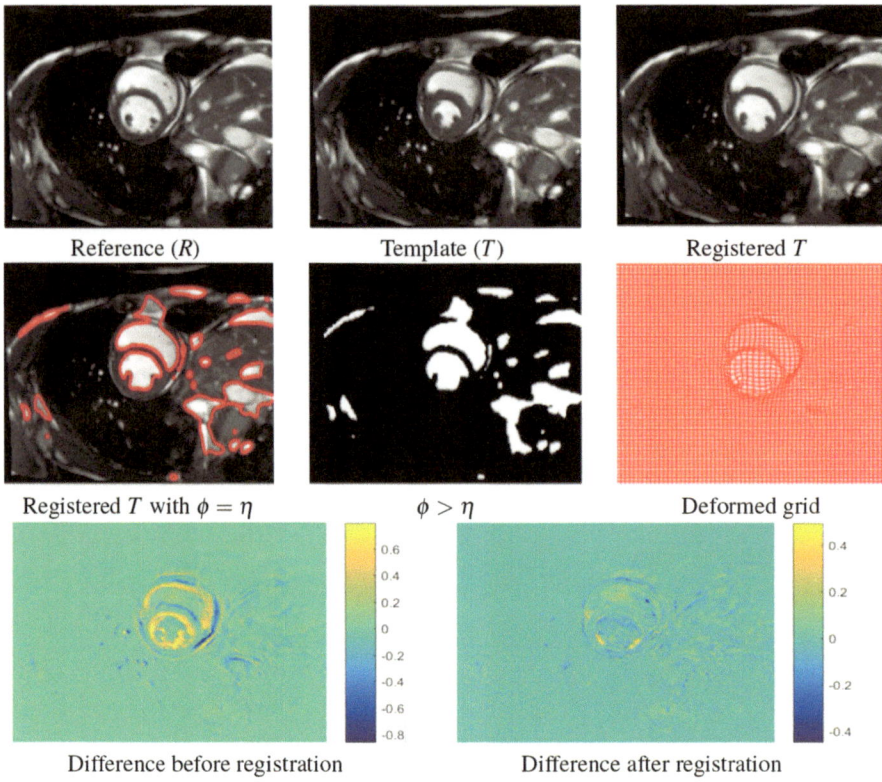

Reference (R)　　　　　　　Template (T)　　　　　　　Registered T

Registered T with $\phi = \eta$　　　　　$\phi > \eta$　　　　　　Deformed grid

Difference before registration　　　　　Difference after registration

Fig. 7.4 The reference and template images are taken from the diastolic and systolic phase of the cardiac cycle. The view plane is the short axis view

Fig. 7.5 This figure shows the convergence behavior of the algorithm on the experiment in Figure 7.4. The total computation time is 6.88 seconds

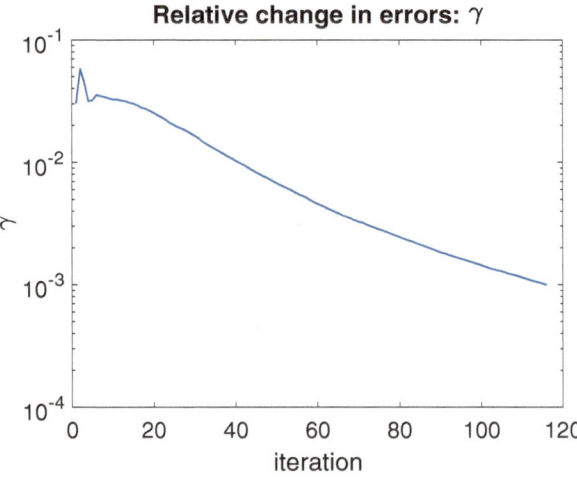

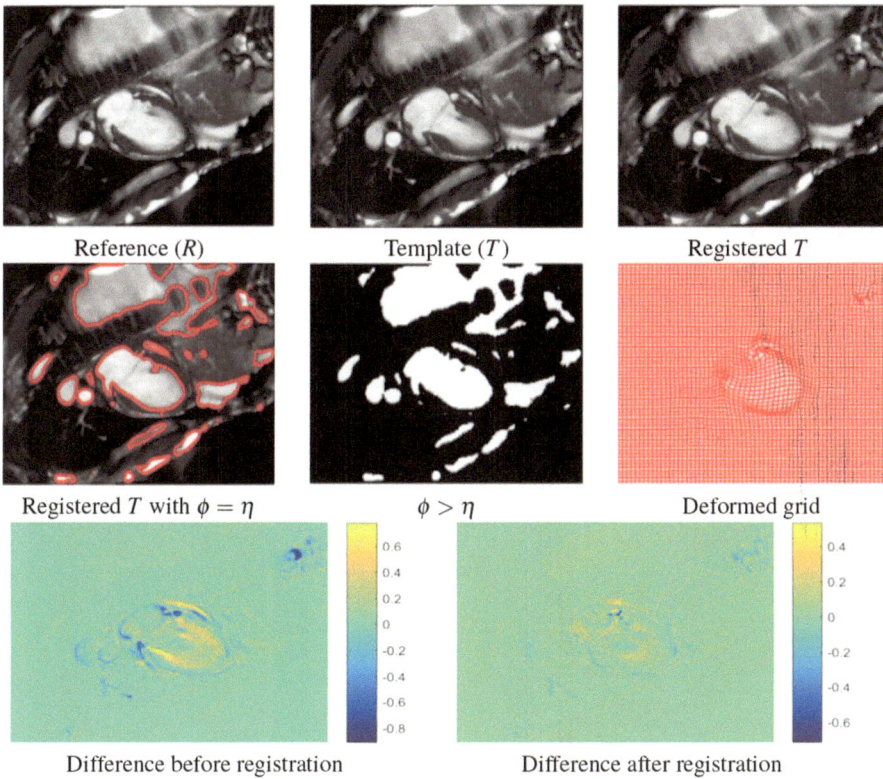

Reference (*R*)   Template (*T*)   Registered *T*

Registered *T* with $\phi = \eta$ $\phi > \eta$ Deformed grid

Difference before registration   Difference after registration

Fig. 7.6 This figure shows results of the algorithm on diastolic and systolic frames of cine images viewed from the vertical long axis view

Fig. 7.7 This figure shows the convergence behavior of the algorithm on the experiment in Figure 7.6. The total computation time is 4.25 seconds

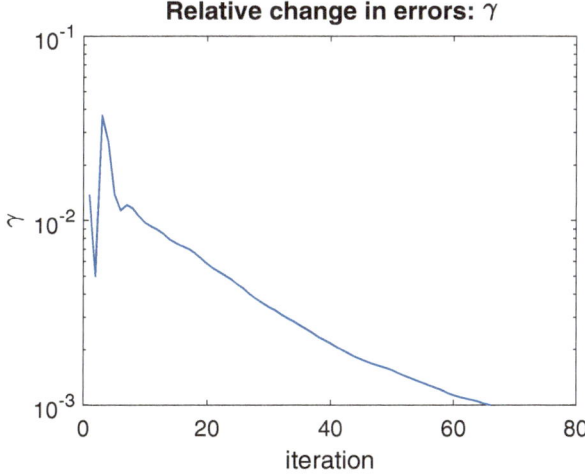

Relative change in errors: γ

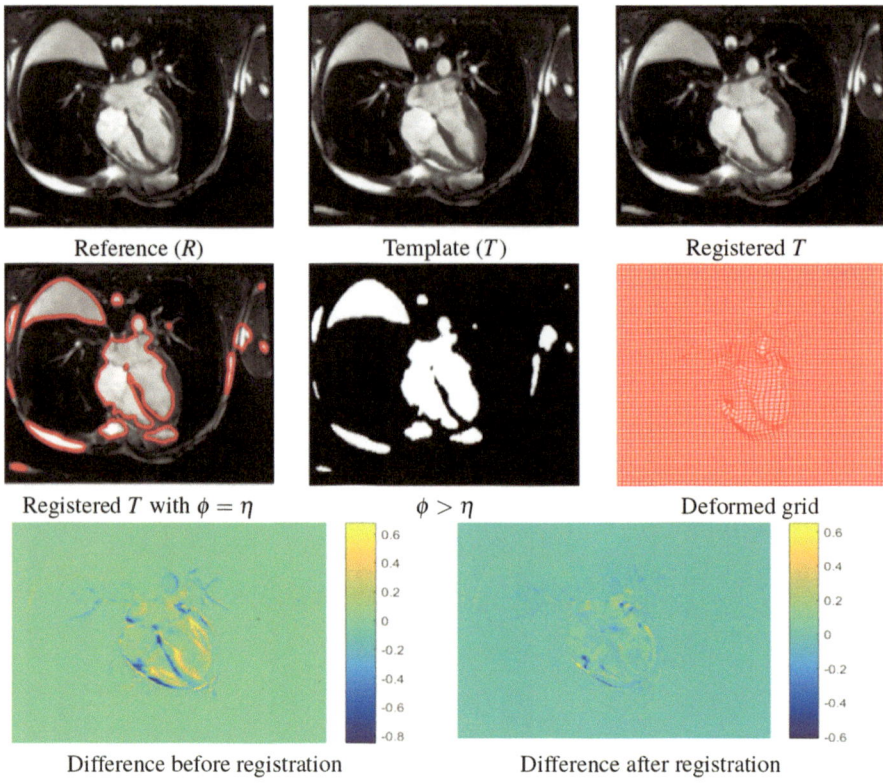

Reference (R)   Template (T)   Registered T

Registered T with $\phi = \eta$ $\phi > \eta$ Deformed grid

Difference before registration   Difference after registration

Fig. 7.8   Another results on diastolic and systolic images from horizontal long axis view

Fig. 7.9   This figure shows the convergence behavior of the algorithm on the experiment in Figure 7.8. The total computation time is 2.60 seconds

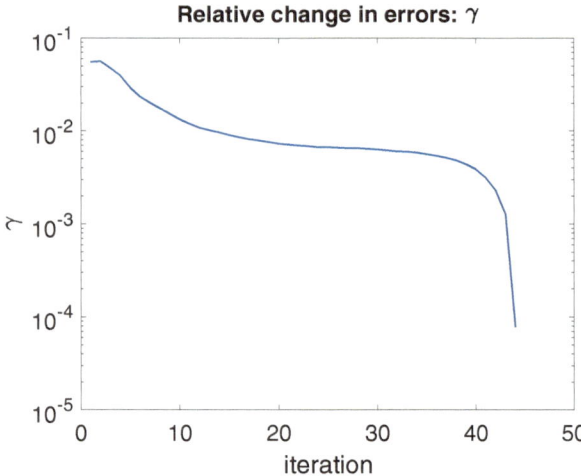

The adopted update scheme allows the segmenting curve to track the changes in the template image during evolution of the displacement vector. Additionally, the semi-implicit discretization allows for stability of the evolution equation for the displacement vectors $\mathbf{u}_1$ and $\mathbf{u}_2$, through implicit treatment of the linear operator in the Euler-Lagrange equations.

As a drawback to the proposed method, we have not integrated mechanisms to account for inherent challenges common in deformable registration such as folding and self-penetration. One way to handle such phenomena is to impose constraints, for instance, only diffeomorphisms are admissible for the registration subproblem. This is a very common constraint in the registration literature and also translates into the following simple mathematical expression:

$$\det(\nabla\psi) = \det(\nabla\mathbf{u} + I) > 0.$$

Integrating this strict diffeomorphic constraint into the model with the foresight of using the fast Fourier transform is not numerically straightforward, and we are currently working on feasible implementations.

Another concern from a theoretical standpoint relates to the non-quasiconvexity of the stored energy density of *St. Venant-Kirchhoff* material models (see [15]). This raises questions on the existence of minimizers for models involving such energies as presented in this paper. From a numerical standpoint, however, experimental results show that sufficient and admissible results can be obtained without the need to finding convex envelopes of the energy as done in [13]. In fact, this observation was first pointed out by Guyader and Vese in [12].

References

1. Ascher, U., Ruuth, S., Wetton, B.: Implicit-explicit methods for time-dependent partial differential equations. SIAM J. Numer. Anal. **32**(3), 797–823 (1995)
2. Aujol, J.F., Gilboa, G., Chan, T.F., Osher, S.: Structure-texture image decomposition - modeling, algorithms, and parameter selection. Int. J. Comput. Vis. **67**(1), 111–136 (2006)
3. Bresson, X., Esedolu, S., Vandergheynst, P., Thiran, J.P., Osher, S.: Fast global minimization of the active contour/snake model. J. Math. Imaging Vis. **28**(2), 151–167 (2007)
4. Chambolle, A.: An algorithm for total variation minimization and applications. J. Math. Imaging Vis. **20**(1), 89–97 (2004)
5. Chan, T.F., Vese, L.A.: Active contours without edges. IEEE Trans. Image Process. **10**(2), 266–277 (2001)
6. Chan, T.F., Golub, G.H., Mulet, P.: A nonlinear primal-dual method for total variation-based image restoration. SIAM J. Sci. Comput. **20**(6), 1964–1977 (1999)
7. Collignon, A., Maes, F., Delaere, D., Vandermeulen, D., Suetens, P., Marchal, G.: Automated multi-modality image registration based on information theory. In: Information Processing in Medical Imaging, vol. 3, pp. 263–274 (1995)
8. Csisz, I., Tusnády, G., et al.: Information geometry and alternating minimization procedures. Stat. Decis. **1**, 205–237 (1984)
9. Davis, P.: Circulant Matrices. John Wiley & Sons, New York (1979)

10. Haber, E., Modersitzki, J.: Intensity gradient based registration and fusion of multi-modal images. In: International Conference on Medical Image Computing and Computer-Assisted Intervention, pp. 726–733. Springer, Berlin (2006)
11. Hansen, P.C., Nagy, J.G., O'leary, D.P.: Deblurring images: matrices, spectra, and filtering. SIAM, Philadelphia, PA (2006)
12. Le Guyader, C., Vese, L.A.: A combined segmentation and registration framework with a nonlinear elasticity smoother. Comput. Vis. Image Underst. **115**(12), 1689–1709 (2011)
13. Ozeré, S., Gout, C., Le Guyader, C.: Joint segmentation/registration model by shape alignment via weighted total variation minimization and nonlinear elasticity. SIAM J. Imaging Sci. **8**(3), 1981–2020 (2015)
14. Pluim, J., Maintz, J., Viergever, M.: Image registration by maximization of combined mutual information and gradient information. In: International Conference on Medical Image Computing and Computer-Assisted Intervention, pp. 452–461. Springer, Berlin (2000)
15. Raoult, A.: Non-polyconvexity of the stored energy function of a Saint Venant-Kirchhoff material. Aplikace Matematiky **31**(6), 417–419 (1986)
16. Viola, P., Iii, W.M.W.: Alignment by maximization of mutual information. Int. J. Comput. Vis. **24**(2), 137–154 (1997)
17. Yezzi, A., Zollei, L., Kapur, T.: A variational framework for joint segmentation and registration. In: Mathematical Methods in Biomedical Image Analysis, pp. 44–51. IEEE, Piscataway, NJ (2001)

Chapter 8
Multi-parameter Mumford-Shah Segmentation

Murat Genctav and Sibel Tari

Abstract Mumford-Shah functional has two parameters that define a two-dimensional scale space of solutions. Instead of using the solution obtained at a predetermined fine-tuned parameter setting, we consider solutions at multiple parameter settings simultaneously. Using multiple solutions, we construct pixel-based features and employ them to extract shapes in images. We experiment with both synthetic and real images.

8.1 Introduction

Mumford and Shah [7] model is one of the widely used segmentation models in image processing. It is based on a minimization of a functional via which a piecewise smooth approximation of a given image (cartoon) and an edge set are to be recovered simultaneously:

$$E_{MS}(u, \Gamma) = \beta \int_R (u - g)^2 dx + \alpha \int_{R \setminus \Gamma} |\nabla u|^2 dx + length(\Gamma) \qquad (8.1)$$

where

- $R \subset \Re^2$ is connected, bounded, and an open subset representing the image domain.
- g is an image defined on R.
- $\Gamma \subset R$ is the edge set segmenting R.
- u is the piecewise smooth approximation of g.
- α, β are the scale space parameters of the model.

M. Genctav (✉)
Middle East Technical University, Ankara, Turkey
e-mail: genctav@ceng.metu.edu.tr

S. Tari
Department of Computer Engineering, Middle East Technical University, Ankara, Turkey
e-mail: stari@metu.edu.tr

© The Author(s) and the Association for Women in Mathematics 2018   133
A. Genctav et al. (eds.), *Research in Shape Analysis*, Association for Women
in Mathematics Series 12, https://doi.org/10.1007/978-3-319-77066-6_8

The first term in $E_{MS}(u, \Gamma)$ is the data-fidelity term, which forces the solution u to be close to the original image g. The other two terms are regularization terms, which encode a priori information about the solution and give preference to piecewise smooth images with simple edge sets. The balance between three competing terms is obtained by the choice of free parameters.

The model inspired many segmentation works. Several computational methods for minimizing the original model or its variants are proposed. Among others, Chan and Vese [3] proposed a level-set based framework for a piecewise constant approximation, El Zehiry et al. [4] proposed a graph cut optimization, and Shah [10] proposed breaking up the functional into coupled equations. The general framework is also extended in several ways to incorporate higher-level information into segmentation process. For example, Erdem and Tari [5] introduced edge linking and hysteresis, Patz et al. [8] extended formulations to uncertain images, and recently Guo et al. [6] combined with sparse dictionary learning to achieve robustness under extreme noise.

Regardless of the method used for minimization and regardless of possible extensions, the models have at least two parameters denoting the smoothing radius and the contrast threshold, two ubiquitous parameters of the formal image processing. The balance between the three competing terms in (8.1), hence the resulting solution is determined by the parameter values. Parameter estimation in the MS setting is not an easy task. Hence, the parameters typically need to be handpicked. There are few papers addressing the issue [2, 9] for special cases.

In this paper, instead of trying to choose the best parameter values, we explore the possibility of integrating multiple solutions obtained by exploring the multi-parameter solution space. Instead of directly handling the original functional, we employ its approximation replacing Γ with a smooth indicator function v defined over the entire image domain. The unknown edge set Γ of a lower dimension makes the minimization difficult, and a convenient approximation is obtained via Γ-convergence framework [1].

We treat the multiple solutions as components of a feature vector which are to be projected into a lower-dimensional space. This way, we obtain a rich description of the image which obviates the need for choosing the appropriate parameters of the Mumford-Shah model.

8.2 Computing u and v

The idea is to define a smooth indicator function $v_\rho : R \rightarrow (0, 1)$. It depends on a parameter ρ, and as $\rho \rightarrow 0$, $v \rightarrow 1 - \chi_\Gamma$. That is, $v(x) \approx 0$ if $x \in \Gamma$ and $v(x) \approx 1$ otherwise. The cardinality of the edge set Γ can be approximated by $\frac{1}{2}\left(\rho|\nabla v|^2 + \frac{(1-v)^2}{\rho}\right)$. Hence, the Mumford-Shah functional in (8.1) is approximated by

$$E_{AT}(u, v) = \int_R \left(\beta(u - g)^2 + \alpha(v^2|\nabla u|^2) + \frac{1}{2}\left(\rho|\nabla v|^2 + \frac{(1 - v)^2}{\rho}\right) \right) dx \quad (8.2)$$

The most appealing property of the approximation (8.2) is that one can apply gradient descent to obtain the condition for minima in the form of a system of coupled PDEs:

$$\frac{\partial u}{\partial t} = \nabla \cdot (v^2 \nabla u) - \frac{\beta}{\alpha}(u - g); \qquad \left.\frac{\partial u}{\partial n}\right|_{\partial R} = 0 \qquad (8.3)$$

$$\frac{\partial v}{\partial t} = \nabla^2 v - \frac{2\alpha |\nabla u|^2 v}{\rho} - \frac{(v - 1)}{\rho^2}; \qquad \left.\frac{\partial v}{\partial n}\right|_{\partial R} = 0 \qquad (8.4)$$

where ∂R denotes the boundary of R and n denotes the direction normal to ∂R.

The coupled equations can be simultaneously solved for u and v using standard numerical discretization techniques such as finite differences. When these equations are discretized using a modified explicit scheme, the iterations converge in the sense that the rate of change is smaller than a threshold. In each iteration, only one variable is updated, while the other variable is kept fixed.

8.3 Feature Space Construction and Its Spectral Exploration

We obtain multiple u and v functions by varying the smoothing radius σ and contrast threshold λ parameters. Once we obtain multiple u and v functions, we construct feature vectors at each image pixel. The actual parameter values α and β are calculated by setting ρ to a small value and then using the relations $\lambda = \sqrt{1/(2\alpha\rho)}$ and $\sigma = \sqrt{2\alpha/\beta}$.

There may be several different strategies in constructing a feature vector for each image pixel. In the simplest case, the feature vector may consist of the multiple function values obtained at the pixel. One may use either or both of the u and v function values. It is also possible to add derivatives, values in neighboring pixels or their statistics.

Our approach to extracting useful information from this high-dimensional representation is to perform dimensionality reduction algorithms. This way, the feature space is transformed in such a way that the first few bases cover most of the intrinsic information contained in the high-dimensional feature space.

8.4 Illustration of the Parameter Space via a Synthetic Image

To demonstrate the scale space behavior of (8.2), we synthesized a grayscale image that involves circles of various sizes and contrasts with the background. The white and the black circles all have the same greater contrast to the background as

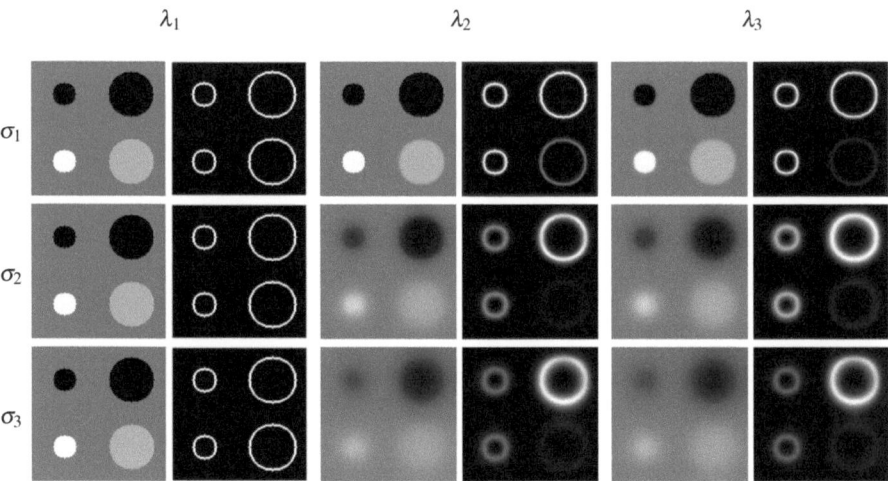

Fig. 8.1 $\{u, 1 - v\}$ function pairs that minimize AT energy given the grayscale image of circles varying the contrast and smoothing radius parameters: $\lambda_i = \{2, 20, 40\}$, $\sigma_i = \{2, 10, 15\}$, $i = 1, 2, 3$. Instead of v function itself, we illustrate $1 - v$ (the edge strength map)

compared to the gray one. $\{u, v\}$ function pairs that result from variation of λ and σ parameters are shown in Figure 8.1. Observe that smoothing is blocked on the edges for the selection of the contrast parameter value λ_1 being quite small compared to any gradient magnitude on those edges.

The second choice for the contrast parameter (λ_2) makes difference. For a smoothing radius (σ_1), which is small compared to the radius of small circles, diffusion is allowed to various extents according to the contrast on the edges. The gray circle with a lower contrast to background is separated from the group of black and white circles that have the same high contrast. This is best observable in $1 - v$, which we will call the edge strength map. In this map, the edge contour of the gray circle is attenuated, whereas black and white circles have edge strength without noticeable difference. However, for larger values of smoothing radius (σ_2 and σ_3), which are comparable to the radii of small and large circles, the edge contour of the bigger black circle stands out in the group of circles having larger contrast.

Finally, the results in the last row/column seem to be analogous but exaggerated versions of their correspondents in the second row/column. Thus, one may consider to select the four $\{u, v\}$ pairs obtained using different parameter configurations from $\{\lambda_1, \lambda_2\} \times \{\sigma_1, \sigma_2\}$ as compact representatives of the two-dimensional scale space, which encapsulate all the information needed for a semantical grouping of the circle objects with respect to their contrast and size.

8.5 Experiments and Discussion

8.5.1 Contrast/Size Experiments

Several experiments are conducted using the results shown in Figure 8.1. Firstly, we represent a node by the function values assigned to that node in a collection of v functions. Then projections of the nodes onto the first three principal components are obtained. Note that at least 80% of the distribution is captured by the first components, which represent an (almost equally weighted) sum of the functions in the collection. In Figure 8.2 we depict color composites formed by first three components. Each column depicts a composite obtained by exploring different subsets of the parameter space.

In columns (a) and (c), the collections consist of the v functions obtained by varying contrast parameter λ given a fixed smoothing radius σ. Using a small smoothing radius (σ_1), we observe that the projections group the circles according to their contrast (a). The white and black circles, which have the same contrast with the background, are grouped together isolating the gray circle with smaller contrast. With a larger smoothing radius (σ_2), the size criterion arises as another distinguishing factor which further separates the large black circle from the group of circles with the same contrast (c). A similar result to the case with column (c) appears in column (b), where only the smoothing radius varies, except the boundary of the gray circle is smoothed due to the selected contrast threshold value (λ_2) which is large compared to the contrast of the gray circle.

In the second experiment, we represent a node by the function values within its neighborhood using a 3×3 window introducing the spatial relationships between nearby nodes. Analogous to the former results without spatial relations introduced, the first principal component takes the sum of the v collection. The spatial relations, however, affect the second and successive components. For example, the second and third components behave like a Prewitt operator that takes the first derivatives in horizontal and vertical directions, respectively. Indeed, these principal components

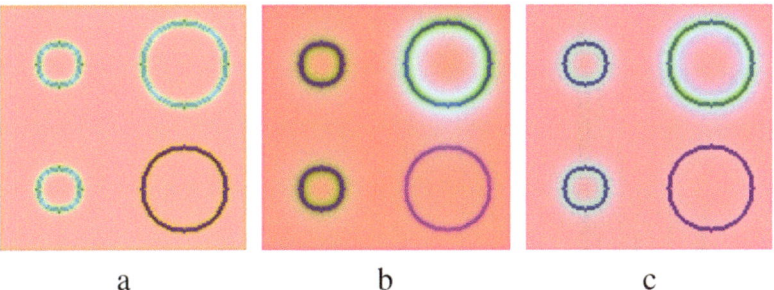

a   b   c

Fig. 8.2 Contrast/size experiments using v. (**a**) Changing the contrast threshold: $(\lambda, \sigma) \in \{\lambda_1, \lambda_2, \lambda_3\} \times \{\sigma_1\}$. (**b**) Changing the smoothing radius: $(\lambda, \sigma) \in \{\lambda_2\} \times \{\sigma_1, \sigma_2, \sigma_3\}$. (**c**) Changing the contrast threshold at a higher smoothing level: $(\lambda, \sigma) \in \{\lambda_1, \lambda_2, \lambda_3\} \times \{\sigma_2\}$

filter the individual v functions using 3×3 Prewitt kernel and take a weighted sum of filter responses. The other projections appear to be just other weighted sums of smoothed v functions.

The next two experiments repeat the first two using u functions instead of v. Not surprisingly, the first projections are almost equally weighted sums of the individual functions. An interesting behavior, on the other hand, is observed in the second projections. They act like a contrast-enhancing difference of Gaussian filter. This behavior, in fact, emerges as an obvious result of the fact that u functions are nothing but Gaussian blurred images with locally adapted blurring radius. What the second components do is just calculating a weighted sum of the Gaussian blurred images using weights that have both positive and negative values. This behavior is observed both when using a u value at a single location and when using 3×3 spatial windows in constructing the feature space. In the latter case, the directional edge detecting filter behavior appears again in various projections.

The color composites formed by the first three projections of the feature space constructed without using spatial windows are shown in Figure 8.3. In Figure 8.4, an illustrative case using spatial windows is presented.

The last experiment deals with the variation in the object-background contrast. We generated the input image in Figure 8.5 (left), where the background exhibits high intensity nonuniformity, and computed *nine* different v-functions using (λ, σ) parameter pairs from the set $\{2, 20, 40\} \times \{2, 10, 15\}$. We depicted the projection

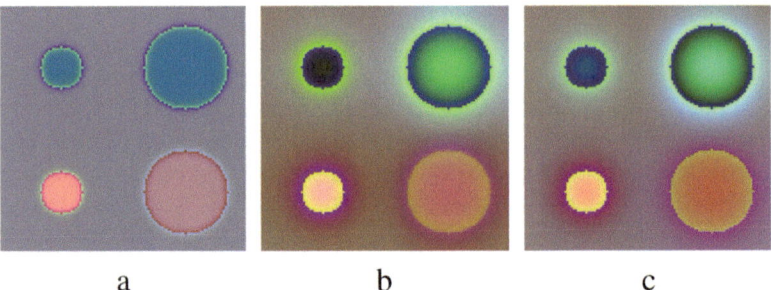

        a   b   c

Fig. 8.3 Contrast/size experiments using u. (**a**) Changing the contrast threshold: $(\lambda, \sigma) \in \{\lambda_1, \lambda_2, \lambda_3\} \times \{\sigma_1\}$. (**b**) Changing the smoothing radius: $(\lambda, \sigma) \in \{\lambda_2\} \times \{\sigma_1, \sigma_2, \sigma_3\}$. (**c**) Changing the contrast threshold at a higher smoothing level: $(\lambda, \sigma) \in \{\lambda_1, \lambda_2, \lambda_3\} \times \{\sigma_2\}$

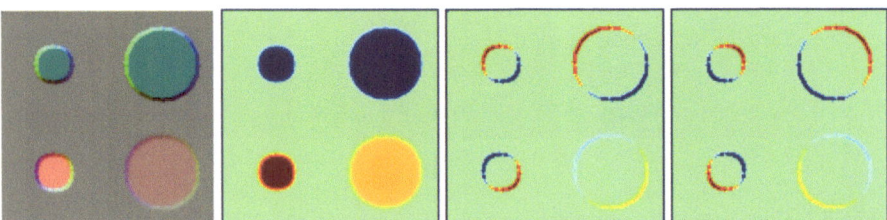

Fig. 8.4 Using u values in a window. $(\lambda, \sigma) \in \{\lambda_1, \lambda_2, \lambda_3\} \times \{\sigma_1\}$

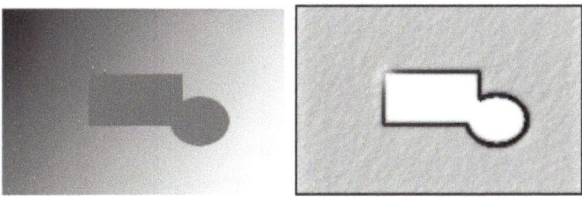

Fig. 8.5 Experiment with an image with high intensity nonuniformity (left). The feature vector is *nine* dimensional – v-values from *nine* different parameter pairs (λ, σ) from the set $\{2, 20, 40\} \times \{2, 10, 15\}$ are concatenated. The right image represents the projection onto the first principal component

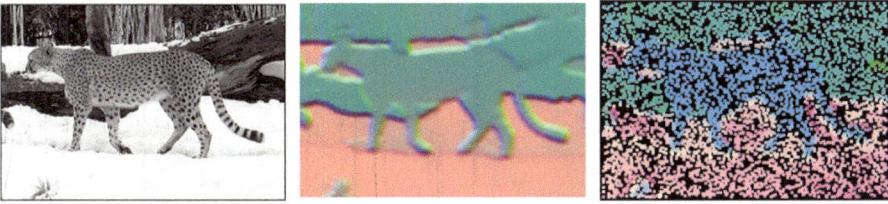

Fig. 8.6 Experiment with cheetah image (left). The feature vector is $26 \times 7 \times 7$ dimensional – u-values from 26 different parameter selections and within a 7×7 spatial window are concatenated. The middle and rightmost images represent the RGB images constructed using first three projections resulting from principal component analysis and local linear embedding, respectively

onto the first component next to the input image. Obviously, the method successfully suppress the variation in the background intensity laying the emphasis on the foreground object without requiring the user to select a perfect pair of parameters.

8.5.2 Complex Images

We experimented with the complex cheetah image depicted in Figure 8.6 using u-function values obtained at 26 different parameter settings where both the contrast threshold and the smoothing radius vary. Each image pixel is represented by a spatial neighborhood of 7×7 window centered at that pixel resulting into a $26 \times 7 \times 7$-dimensional feature vector. We employed principal component analysis and local linear embedding methods for dimensionality reduction and compared RGB images composed of the resulting first three projections in Figure 8.6. Due to the high memory and time requirement of the local linear embedding method, we had to randomly sample pixels from the image to be able to experiment with this method in contrast to the experiment with principal component analysis where all the pixels are used. Clearly, local linear embedding method could better isolate the cheetah from the background (see blue colored points in Figure 8.6 rightmost image).

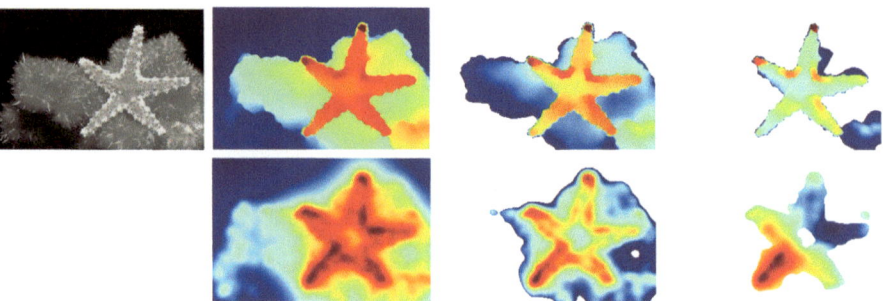

Fig. 8.7 A starfish image (top left) and its iterative segmentations via zero-level sets of the first projections obtained using u-values (top row) and Gabor responses (bottom row) for comparison. At each iteration, the pixel set is split into two based on the sign of the first projection. The u-feature vector is 4 dimensional, whereas the Gabor feature vector is 20 dimensional

We use the complex starfish image depicted in Figure 8.7 to further explore whether we can iteratively use the projections to segment the complex image into meaningful regions. We constructed a *four*-dimensional feature space using u values obtained at *four* different parameter settings where both the contrast threshold and the smoothing radius vary. In Figure 8.7 top row, next to the starfish image, we depict the projections onto the first principal component at successive iterations. At each iteration, the image pixels are splitted into two sets according to the sign of the first projection scores. The background segment (white regions in the visualizations) is excluded in the next iteration.

With the intent of a comparative evaluation, we repeated the last experiment this time using Gabor filter responses as textural features. Note that texture appears to be a significant visual clue that helps separating starfish from the background. To this end, we designated an array of Gabor filters tuned to different orientations ($0°$, $45°$, $90°$, $135°$) and wavelengths (2.82, 5.66, 11.31, 22.63, 45.25) yielding 20-dimensional feature vectors to represent image pixels. In order to compensate for local variations in filter responses, each filter response is smoothed using Gaussian low-pass filters having sigma values selected proportional to the corresponding Gabor wavelength. In Figure 8.7 bottom row, we depicted the resulting projections onto first principal components in the first three iterations.

We start discussing the results with our proposed approach (using u-values) and using texture features (using Gabor responses) with the first iteration (leftmost projections in Figure 8.7). We observe that our approach better localizes the starfish, the surface, and the dark background, whereas with the Gabor features, the segments corresponding to the foreground objects tend to overspread due to Gaussian smoothing. Here, we note that without Gaussian smoothing the results become noisy. If we compare the final results, we see that neither approach surpasses the other. Yet, we observe that our result could be improved with a more sophisticated thresholding method since the star segment shows a significant contrast to the background pixels.

8.6 Summary and Conclusion

We proposed exploiting the multi-parameter solution space of Mumford-Shah image segmentation model instead of relying on a single solution at a fine-tuned parameter setting. Using multiple solutions to a convenient approximation to the model via Γ-convergence framework, we construct a high-dimensional feature vector that is, then, projected into a lower-dimensional space which accounts for as much of the variability in the original space as possible. We demonstrated the capability of the method to distinguish between different appearances of the same shape within the same image with respect to its size and contrast with the background. Then, we showed the effect of introducing spatial relationships via addition of neighboring values in the solutions to the feature vector. We also made an experiment with an image where the object-to-background contrast varies. Finally, we presented methods that use multi-parameter solutions to Mumford-Shah model to successfully segment a target shape from a complex real image.

References

1. Ambrosio, L., Tortorelli, V.M.: Approximation of functional depending on jumps by elliptic functional via t-convergence. Commun. Pure Appl. Math. **43**(8), 999–1036 (1990)
2. Bergounioux, M., Vicente, D.: Parameter selection in a Mumford–Shah geometrical model for the detection of thin structures. Acta Appl. Math. **141**(1), 17–47 (2016)
3. Chan, T.F., Vese, L.A.: Active contours without edges. IEEE Trans. Image Process. **10**(2), 266–277 (2001)
4. El-Zehiry, N., Sahoo, P., Elmaghraby, A.: Combinatorial Optimization of the piecewise constant Mumford-Shah functional with application to scalar/vector valued and volumetric image segmentation. Image Vis. Comput. **29**(6), 365–381 (2011)
5. Erdem, E., Tari, S.: Mumford-Shah regularizer with contextual feedback. J. Math. Imaging Vis. **33**(1), 67–84 (2009)
6. Guo, W., Qin, J., Tari, S.: Automatic prior shape selection for image segmentation. Techreport, Computational and Applied Mathematics, UCLA Department of Mathematics (2014)
7. Mumford, D., Shah, J.: Optimal approximations by piecewise smooth functions and associated variational problems. Commun. Pure Appl. Math. **42**(5), 577–685 (1989)
8. Pätz, T., Preusser, T.: Ambrosio-Tortorelli segmentation of stochastic images. In: ECCV, pp. 254–267. Springer, Berlin (2010)
9. Shah, J.: Parameter estimation, multiscale representation and algorithms for energy-minimizing segmentations. In: ICPR, pp. 815–819, vol. I. (1990)
10. Shah, J.: Segmentation by nonlinear diffusion. In: CVPR, pp. 202–207 (1991)

Chapter 9
L^1-Regularized Inverse Problems for Image Deblurring via Bound- and Equality-Constrained Optimization

Johnathan M. Bardsley and Marylesa Howard

Abstract Image deblurring is typically modeled as an ill-posed, linear inverse problem. By adding an L^1-penalty to the negative-log likelihood function, the resulting minimization problem becomes well-posed. Moreover, the penalty enforces sparsity. The difficulty with L^1-penalties, however, is that they are non-differentiable. Here we replace the L^1-penalty by a linear penalty together with bound and equality constraints. We consider two statistical models for measurement error: Gaussian and Poisson. In either case, we obtain a bound- and equality-constrained minimization problem, which we solve using an iterative augmented Lagrangian (AL) method. Each iteration of the AL method requires the solution of a bound-constrained minimization problem, which is convex-quadratic in the Gaussian case and convex in the Poisson case. We recommend two highly efficient methods for the solution of these subproblems that allows us to apply the AL method to large-scale imaging examples. Results are shown on synthetic data in one and two dimensions, as well as on a radiograph used to calibrate the transmission curve of a pulsed-power X-ray source at a US Department of Energy radiography facility.

9.1 Introduction

In applications such as astronomy, medicine, physics and biology, digital images are used to answer important scientific questions. Imperfections in the underlying model and imaging system cause degradation, or blurring, of recorded images. When such is the case, computational post processing techniques, called image deblurring, are often used.

J. M. Bardsley
The University of Montana, Department of Mathematics, Missoula, MT, USA
e-mail: bardsleyj@mso.umt.edu

M. Howard (✉)
Nevada National Security Site, Las Vegas, NV, USA
e-mail: howardmm@nv.doe.gov

© The Author(s) and the Association for Women in Mathematics 2018
A. Genctav et al. (eds.), *Research in Shape Analysis*, Association for Women in Mathematics Series 12, https://doi.org/10.1007/978-3-319-77066-6_9

Image deblurring is typically modeled as a linear inverse problem. Suppose $f(t)$, $t \in \mathbb{R}^d$, is a function describing the true d-dimensional image, where for us $d = 1$ or 2. The mathematical model of image formation is given by the integral equation

$$d(s) = \int_\Omega k(s, t) f(t) dt,$$

where $s, t \in \mathbb{R}^d$; $d(s)$ is a function that represents the observed image; $\Omega \subset \mathbb{R}^d$ is the computational domain; and the kernel $k(s, t)$ is a function that specifies how the point sources in the image are distorted and is therefore called the point spread function (PSF). The inverse problem of image deblurring is to estimate f given k and d. If the PSF has the property that $k(s, t) = k(s - t)$, which we will assume in our examples, then it is said to be spatially invariant, and the integral operator above takes convolution form: $d(s) = \int_\Omega k(s - t) f(t) dt$. Thus the corresponding inverse problem is called *deconvolution*.

In a realistic problem, the image and PSF are collected only at discrete points within a finite bounded region Ω. It is, therefore, typical to work with a discretization of the above integral equation of the form

$$d = Kf,$$

where $K \in \mathbb{R}^{n \times n}$ is the blurring matrix, $f \in \mathbb{R}^n$ is the unknown image, and $d \in \mathbb{R}^n$ is the observation vector. Note that for two-dimensional images, columns are stacked (sometimes called lexicographical ordering) to obtain the vectors f and d, and K is defined accordingly (see [16] for details).

We consider two measurement error models: Gaussian and Poisson. In the Gaussian case, the measurement model has the form

$$d = Kf + \varepsilon, \tag{9.1}$$

where ε is assumed to be independent and identically distributed Gaussian, which we denote $\varepsilon \sim \mathcal{N}(0, \sigma^2 I)$. In the Poisson case, the measurement model takes the form

$$d = \text{Poiss}(Kf + g), \tag{9.2}$$

where g is the $n \times 1$ vector of background counts and is assumed to be known.

Our task is to estimate f, given a measurement model, a measurement vector d, and the blurring matrix K. A standard approach is to compute the maximum likelihood estimator or, equivalently, the minimum of the log-likelihood. In addition, since f denotes intensities, we add a nonnegativity constraint to obtain

$$\hat{f} = \arg\min_{f \geq 0} \ell(f|d), \tag{9.3}$$

where $\ell(\cdot|d)$ is the log-likelihood, for which, in the Gaussian case, the log-likelihood is defined to be

$$\ell(f|d) = \tfrac{1}{2\sigma^2} \, \|Kf - d\|_2^2 \,, \tag{9.4}$$

and in the Poisson case

$$\ell(f|d) = \sum_{j=1}^{n} \left\{ ([Kf]_j + g_j) - d_j \ln([Kf]_j + g_j) \right\}. \tag{9.5}$$

Although the addition of the nonnegativity constraint in (9.3) serves to stabilize the optimization problem, it remains ill-posed, and so regularization is needed. We focus on regularization methods that involve the L^1-norm. The first type of regularization involves the addition of an equality constraint; specifically, assuming K preserves energy, i.e., $K^T 1 = 1$, and that $f, d \geq 0$, it can be shown that then $\|d\|_1 = \|Kf\|_1 = \|f\|_1$. Adding this constraint to (9.3) yields

$$\hat{f} = \arg\min_{f \geq 0} \ell(f|d) \quad \text{s.t.} \quad 1'f = \|d\|_1, \tag{9.6}$$

where $1 \in \mathbb{R}^n$ denotes the constant vector of 1's and $\ell(f|d)$ is given either by (9.4) or (9.5). We call this the regularization method *energy preservation* in what follows.

A more standard regularization method is to add a penalty term to the objective function, i.e.,

$$\hat{f} = \arg\min_{f \geq 0} \{\ell(f|d) + \alpha J(f)\}, \tag{9.7}$$

where $\alpha > 0$ is the regularization parameter, the choice of which we do not discuss here, but can be found in [15, 16]. Standard choices for the regularization function J include the quadratic functions $J(f) = \|f\|_2^2$ and $J(f) = \|Df\|_2^2$, where D is a discretization of the gradient operator. In this paper, we focus on the non-quadratic regularization functions that arise when the L^1-norm is used. These include $J(f) = \|f\|_1$, and $J(f) = \|Df\|_1$, where D denotes either the discrete gradient (yielding total variation (TV) regularization) or the discrete wavelet transform (yielding Besov regularization).

When an L^1-penalty function is used, the nondifferentiability of the absolute function presents a problem for gradient-based optimization methods. This has been overcome in previous work (see, e.g., [16]) by replacing the absolute value with a differentiable approximation, such as $|t| \approx \sqrt{t^2 + \beta}$, for $0 < \beta << 1$. A more recently developed approach for solving optimization problems of the form (9.7), where $J(f)$ is the TV penalty, is exemplified by Beck and Teboulle [5], which makes use of the "proximal map" and the gradient projection method to solve the dual optimization problem. Perhaps the most popular current approach is the so-called alternating direction method of multipliers (ADMM), where (9.7) is replaced with an equivalent optimization problem in which the absolute value function does not appear, but auxiliary variables and additional constraints are added. For example, in the ADMM algorithm of [7], auxiliary variables are introduced in order to address both the nondifferentiability of the TV penalty and the nonnegativity constraints on the unknown f. An augmented Lagrangian is then minimized using an alternating direction method, yielding a solution to (9.7). In this paper, we instead formulate our optimization problems as in [11], using auxiliary variables to overcome the nondifferentiability of the TV penalty and a bound-constrained optimization method to deal with the nonnegativity constraints. Moreover, in our approach, all of the unknowns are optimized simultaneously rather than by an alternating direction algorithm. More specifically, we take the general augmented Lagrangian (AL) approach described in [12, Algorithm 17.4], though with a simpler (and more effective for this problem) penalty update. Moreover, rather than using the projected gradient algorithm, as is suggested in [12, Algorithm 17.4], we use the much more efficient gradient projection conjugate gradient (GPCG) method [10] in the Gaussian case and the gradient projection-reduced Newton (GPRN) method [3] in the Poisson case. In our experiments, incorporating the nonnegativity constraints directly into the problem, while more computationally demanding, yields better results and algorithms that require less tuning.

The remainder of the paper goes as follows. In Section 9.2, we provide a detailed derivation of the optimization problems that we wish to solve. Then in Section 9.3, we present our optimization algorithms. Numerical experiments are provided in Section 9.4, and we end with conclusions in Section 9.5.

9.2 Bound- and Equality-Constrained Optimization Problems

We begin by representing the L^1 regularization function as a linear function plus constraints, following the ideas in [11]. For one-dimensional TV, and for both one- and two-dimensional Besov regularization, the L^1 regularization function can be written

$$\|\boldsymbol{D}\boldsymbol{f}\|_1 = \left\{ \mathbf{1}^T (\boldsymbol{v}_+ + \boldsymbol{v}_-) \mid \boldsymbol{D}\boldsymbol{f} = \boldsymbol{v}_+ - \boldsymbol{v}_-, \ \& \ \boldsymbol{v}_+, \boldsymbol{v}_- \geq \boldsymbol{0} \right\}. \qquad (9.8)$$

Note that $\boldsymbol{v}_+ = \max\{\boldsymbol{D}\boldsymbol{f}, \boldsymbol{0}\}$ and $\boldsymbol{v}_- = \max\{-\boldsymbol{D}\boldsymbol{f}, \boldsymbol{0}\}$, where the max function is evaluated component-wise. We assume that $\boldsymbol{D}$ is an $n \times n$ matrix, so that $\boldsymbol{f}, \boldsymbol{v}_+, \boldsymbol{v}_- \in \mathbb{R}^n$.

This approach can be extended to the two-dimensional TV case, but the anisotropic TV function must be used:

$$J(\boldsymbol{f}) = \|\boldsymbol{D}_h \boldsymbol{f}\|_1 + \|\boldsymbol{D}_v \boldsymbol{f}\|_1, \qquad (9.9)$$

where $\boldsymbol{D}_h$ and $\boldsymbol{D}_v$ are the $n \times n$ horizontal and vertical derivative matrices, respectively. This regularization function is both computationally and theoretically very similar to standard TV (see the results in [2]). Now, proceeding as above for both $\boldsymbol{D}_h$ and $\boldsymbol{D}_v$, we obtain

$$\|\boldsymbol{D}_h\boldsymbol{f}\|_1 = \left\{\mathbf{1}^T(\boldsymbol{v}_+^h + \boldsymbol{v}_-^h) \mid \boldsymbol{D}_h\boldsymbol{f} = \boldsymbol{v}_+^h - \boldsymbol{v}_-^h, \,\& \, \boldsymbol{v}_+^h, \boldsymbol{v}_-^h \geq \mathbf{0}\right\} \quad (9.10)$$

$$\|\boldsymbol{D}_v\boldsymbol{f}\|_1 = \left\{\mathbf{1}^T(\boldsymbol{v}_+^v + \boldsymbol{v}_-^v) \mid \boldsymbol{D}_v\boldsymbol{f} = \boldsymbol{v}_+^v - \boldsymbol{v}_-^v, \,\& \, \boldsymbol{v}_+^v, \boldsymbol{v}_-^v \geq \mathbf{0}\right\}, \quad (9.11)$$

where $\boldsymbol{v}_+^h$ and $\boldsymbol{v}_-^h$ are defined as above for the directional derivative matrix $\boldsymbol{D}_h$ and $\boldsymbol{v}_+^v$ and $\boldsymbol{v}_-^v$ are defined as above for $\boldsymbol{D}_v$.

Incorporating these representations of the L^1-penalty into (9.7) will yield the optimization problems that will be our focus. In all cases, the optimization problem will take the general form

$$\hat{\boldsymbol{x}} = \arg\min_{\boldsymbol{l} \leq \boldsymbol{x} \leq \boldsymbol{u}} \mathcal{T}_\alpha(\boldsymbol{x}), \quad \text{such that} \quad \boldsymbol{C}\boldsymbol{x} = \boldsymbol{r}, \quad (9.12)$$

where $\boldsymbol{l} = \mathbf{0}$, $\boldsymbol{u} = \infty$; $\mathcal{T}_\alpha$ is either convex-quadratic or simply convex in $\boldsymbol{x}$ and α denotes the regularization parameter. The matrix $\boldsymbol{C}$ and vector $\boldsymbol{r}$ will be determined by which regularization is used: for one-dimensional TV and Besov regularization, $\boldsymbol{x} = (\boldsymbol{f}, \boldsymbol{v}_+, \boldsymbol{v}_-)$, whereas for two-dimensional TV, $\boldsymbol{x} = (\boldsymbol{f}, \boldsymbol{v}_+^h, \boldsymbol{v}_-^h, \boldsymbol{v}_+^v, \boldsymbol{v}_-^v)$. The function $\mathcal{T}_\alpha$, matrix $\boldsymbol{C}$, and vector $\boldsymbol{b}$ for the various cases are described in more detail next.

9.2.1 Gaussian Case: Bound- and Equality-Constrained Quadratic Programs

In the Gaussian case, $\ell(\boldsymbol{f}|\boldsymbol{d})$ is defined by (9.4), where the noise variance σ^2 can be ignored. For the cases of one-dimensional TV regularization or Besov regularization in either one or two dimensions, we construct (9.7) as the optimization problem in (9.12), where $\mathcal{T}_\alpha$ is defined as

$$\mathcal{T}_\alpha(\boldsymbol{x}) = \tfrac{1}{2}\boldsymbol{x}'\boldsymbol{A}\boldsymbol{x} - \boldsymbol{x}'\boldsymbol{b}, \quad (9.13)$$

with $\boldsymbol{x} = (\boldsymbol{f}, \boldsymbol{v}_+, \boldsymbol{v}_-)$,

$$\boldsymbol{A} = \begin{bmatrix} \boldsymbol{K}'\boldsymbol{K} & \mathbf{0} & \mathbf{0} \\ \mathbf{0} & \mathbf{0} & \mathbf{0} \\ \mathbf{0} & \mathbf{0} & \mathbf{0} \end{bmatrix}, \quad \text{and} \quad \boldsymbol{b} = \begin{bmatrix} \boldsymbol{K}'\boldsymbol{d} \\ -\alpha\mathbf{1} \\ -\alpha\mathbf{1} \end{bmatrix}. \quad (9.14)$$

The equality constraint terms in (9.12) are defined as $\boldsymbol{r} = \mathbf{0}$ and $\boldsymbol{C} = [\boldsymbol{D} \ -\boldsymbol{I} \ \boldsymbol{I}]$, with $\boldsymbol{I}$ the identity matrix.

In the two-dimensional TV case, (9.7) may be rewritten in the form of (9.12) with $\mathcal{T}_\alpha$ given by (9.13), where $x = (f, v_+^h, v_-^h, v_+^v, v_-^v)$,

$$
A = \begin{bmatrix} K'K\,0\,0\,0\,0 \\ 0\;\;0\,0\,0\,0 \\ 0\;\;0\,0\,0\,0 \\ 0\;\;0\,0\,0\,0 \\ 0\;\;0\,0\,0\,0 \end{bmatrix}, \quad b = \begin{bmatrix} K'd \\ -\alpha 1 \\ -\alpha 1 \\ -\alpha 1 \\ -\alpha 1 \end{bmatrix}, \tag{9.15}
$$

and

$$
C = \begin{bmatrix} D_h\;-I\;I\;\;0\;\;0 \\ D_v\;\;\;0\;\;0\;-I\;I \end{bmatrix}. \tag{9.16}
$$

Note that although (9.12) and (9.13) do not have differentiability issues, the number of unknowns has increased, from n to $3n$ for one-dimensional TV and for Besov regularization and from n to $5n$ for two-dimensional TV. This is not insignificant, especially in the two-dimensional cases where n is already quite large.

Finally, we note that the optimization problem (9.6) (i.e., the *energy preservation* case) also has the form (9.12) but with $\mathcal{T}_\alpha(x)$ as in (9.13), $x = f$, $A = K'K$, $b = K'd$, $C = 1'$, and $r = \|d\|_1$; thus our optimization method described below may be applied to this problem as well.

9.2.2 Poisson Case: Bound- and Equality-Constrained Convex Programs

In the Poisson noise case, $\ell(f|b)$ is defined by (9.5). To obtain an optimization problem of the form (9.12), beginning with (9.7), for the one-dimensional TV regularization and one- or two-dimensional Besov regularization cases, we define

$$
\mathcal{T}_\alpha(x) = \sum_{j=1}^{n} \left\{ ([Kf]_j + g_j) - d_j \ln([Kf]_j + g_j) \right\} + \alpha 1^T (v_+ + v_-), \tag{9.17}
$$

where $\mathcal{T}_\alpha(x) \overset{def}{=} \mathcal{T}_\alpha(f, v_+, v_-)$ and, as in the Gaussian case, $C = [D\;-I\;I]$, and $r = 0$. For the two-dimensional TV case, (9.7) is rewritten as (9.12) with

$$
\mathcal{T}_\alpha(x) = \sum_{j=1}^{n} \left\{ ([Kf]_j + g_j) - d_j \ln([Kf]_j + g_j) \right\}
$$
$$
+ \alpha 1^T (v_+^h + v_-^h + v_+^v + v_-^v), \tag{9.18}
$$

where $\mathcal{T}_\alpha(x) \overset{def}{=} \mathcal{T}_\alpha(f, v_+^h, v_-^h, v_+^v, v_-^v)$ and, as in the Gaussian case, $r = 0$ and C is defined by (9.16).

Finally, note that in the Poisson case, the optimization problem (9.6) has the form (9.12) with $x = f$, $C = 1'$, and $r = \|d\|_1$; thus our optimization method described below can be applied to this problem as well.

9.3   Augmented Lagrangian Techniques for Solving Problem (9.12)

We have now stated the general form optimization problem that we wish to solve, namely (9.12), where $\mathcal{T}_\alpha$ is a convex (and perhaps quadratic) function. In this section, we present an algorithm for solving such problems, and then we present details specific to either the Gaussian or Poisson cases. We take an augmented Lagrangian (AL) approach [6], which modifies the cost function of (9.12) to include a Lagrange multiplier and a penalty term as follows:

$$\mathcal{L}_\alpha(x, \lambda, \mu) \overset{def}{=} \mathcal{T}_\alpha(x) + \lambda'(Cx - r) + \tfrac{\mu}{2}\|Cx - r\|_2^2, \tag{9.19}$$

where $\lambda \in \mathbb{R}^M$ is the vector of Lagrange multipliers and $\mu > 0$ is a penalty parameter. Note that as μ increases, failure to satisfy the equality constraint $Cx = r$ is increasingly penalized, forcing the optimization toward the feasible region.

We now describe our method, which generally speaking is close to [12, Algorithm 17.4]. At iteration k, given the approximations (x_k, λ_k, μ_k), we first update x by solving

$$x_{k+1} = \arg\min_{l \leq x \leq u} \mathcal{L}_\alpha(x, \lambda_k, \mu_k), \tag{9.20}$$

where

$$\mathcal{L}_\alpha(x, \lambda_k, \mu_k) = \mathcal{T}_\alpha(x) + \lambda_k'(Cx - r) + \tfrac{\mu_k}{2}\|Cx - r\|_2^2, \tag{9.21}$$

with $\mathcal{T}_\alpha$ defined as (9.13), (9.14) or (9.13), (9.15) in the Gaussian case and as (9.17) or (9.18) in the Poisson case. To solve (9.20), (9.21) in the Gaussian case, where $\mathcal{T}_\alpha$ is quadratic, we advocate using the gradient projection conjugate gradient algorithm (GPCG) [10], which is very efficient for such problems. In the Poisson case, on the other hand, we advocate the use of the gradient projection-reduced Newton (GPRN) algorithm of [3] to solve (9.20), (9.21); GPRN was designed specifically for the minimization of the negative-log Poisson likelihood plus a penalty function and a nonnegativity constraint and is very efficient for such problems.

Both GPCG and GPRN are theoretically convergent algorithms for the problems we present here [1, 3, 10]. The two algorithms have the same structure, interspersing gradient projection iterations, which are effective for active set identification, with conjugate gradient iterations, which significantly accelerate convergence over gradient projection iterations alone. For more details on these algorithms, as well as comparisons with other methods, see [1, 3, 10]. If $\{x_{k,j}\}$ are the GPCG or GPRN iterates, indexed on j, we stop iterations once either the relative projected gradient norm or the relative step size norm goes below a predetermined stopping tolerance, e.g., 10^{-6}. The relative step size norm is defined $\|x_{k,j+1} - x_{k,j}\|/\|x_{k,0}\|$, while the relative projected gradient norm is defined $\|\nabla_{\mathcal{P}}\mathcal{L}_\alpha(x_{k,j+1}, \lambda_k, \mu_k)\|/\|\nabla_{\mathcal{P}}\mathcal{L}_\alpha(x_{k,0}, \lambda_k, \mu_k)\|$, where the projected gradient is defined

$$[\nabla_{\mathcal{P}}\mathcal{L}_\alpha(x, \lambda_k, \mu_k)]_i = \begin{cases} [\nabla_x\mathcal{L}_\alpha(x, \lambda_k, \mu_k)]_i, & \text{if } x_i \in (l_i, u_i), \\ \min\{[\nabla_x\mathcal{L}_\alpha(x, \lambda_k, \mu_k)]_i, 0\}, & \text{if } x_i = l_i, \\ \max\{[\nabla_x\mathcal{L}_\alpha(x, \lambda_k, \mu_k)]_i, 0\}, & \text{if } x_i = u_i. \end{cases}$$

A maximum number of GPCG or GPRN iterations may be set for additional termination criteria, and in all examples that follow, the algorithms terminated in no more than 30 iterations.

With the estimate x_{k+1} in hand, we then update λ_k. To derive the update equation for λ_k, we note that the solution (x^*, λ^*, μ^*) satisfies

$$0 = \nabla_x\mathcal{L}_\alpha(x^*, \lambda^*, \mu^*)$$

$$= \nabla_x\mathcal{T}_\alpha(x^*) + C'\lambda^* + \mu^*C'(Cx^* - r) \tag{9.22}$$

$$= \nabla_x\mathcal{T}_\alpha(x^*) + C'\lambda^*, \tag{9.23}$$

since $Cx^* = r$. Moreover, at iterate $(x_{k+1}, \lambda_k, \mu_k)$, considering Equation (9.22), we have

$$0 \approx \nabla_x\mathcal{T}_\alpha(x_{k+1}) + C'(\lambda_k + \mu_k(Cx_{k+1} - r)). \tag{9.24}$$

Finally, comparing (9.23) and (9.24), we obtain the update formula

$$\lambda_{k+1} = \lambda_k + \mu_k(Cx_{k+1} - r). \tag{9.25}$$

Various updates for penalty parameter μ_k are given in [6], and one also appears in [12, Algorithm 17.4]. We use one from [6], which is straightforward to implement and works well for our problems: let $\beta > 1$ and $0 < \nu < 1$, then

$$\mu_{k+1} = \begin{cases} \beta\mu_k & \text{if } \|Cx_{k+1} - r\| > \nu\|Cx_k - r\|, \\ \mu_k, & \text{if } \|Cx_{k+1} - r\| \leq \nu\|Cx_k - r\|. \end{cases} \tag{9.26}$$

This increases the penalty parameter by a factor of β only if the constraint violation has not decreased by a factor of ν from the previous iteration. In our experience, β must be chosen carefully: if β is too large, μ_k increases too rapidly, and the approximate solution x_k becomes poor before convergence; however, if β is too small, μ_k increases too slowly, and the AL method will be slow to converge. In all of the applications that we consider, we set $\beta = 2$ and $\nu = 1/4$, and we found that the choice of μ_0 needs to be tuned somewhat for different cases.

Combining (9.20), (9.25), and (9.26) yields our AL method for solving the optimization problem (9.12). Pseudo-code for the AL method is presented in Algorithm 9.1. Convergence of the method is proved in [6, 12].

Algorithm 9.1 The Bound-Constrained Augmented Lagrangian Method

---

0. Given $\mu_0 \geq 0$, x_0, $\lambda_0 = 1$, and $0 < \varepsilon << 1$, set $k = 0$.
1. Use GPCG in the Gaussian noise case and GPRN in the Poisson noise case to solve (9.20), stopping iterations once the relative projected gradient norm or relative step norm tolerance has been met or the maximum GPCG/GPRN iterations has been reached. Define x_{k+1} to be the final iterate.
2. Define $\rho_{k+1} := \|Cx_{k+1} - r\|/\|Cx_0 - r\|$ and stop iterations once $\rho_{k+1} < \varepsilon$ or $(\rho_{k+1} - \rho_k)/\rho_k < \varepsilon$ with approximate solution x_{k+1}.
3. Update the Lagrange multiplier via (9.25) to obtain λ_{k+1}, and choose a new penalty parameter μ_{k+1} via (9.26).
4. Set $k = k + 1$ and go to Step 1.

9.4 Numerical Results

We are now ready to test our AL method on both one- and two-dimensional test cases.

9.4.1 One-Dimensional Synthetic Image Deconvolution

We begin with a one-dimensional deblurring example. The data model is of the form (9.1), with $d = Kf$ obtained via mid-point quadrature applied to the convolution equation

$$d(s) = \int_0^1 K(s - s')f(s')ds',$$

where $K(s) = \exp(-s^2/(2\gamma^2))/\sqrt{\pi\gamma^2}$, $\gamma > 0$. Then K has the form

$$[K]_{ij} = h \exp\left(-((i - j)h)^2/(2\gamma^2)\right)/\sqrt{\pi\gamma^2}, \quad 1 \leq i, j \leq n,$$

where $h = 1/n$ with n the number of grid points in $[0, 1]$. We use $n = 256$, and the resulting K has full column rank with condition number on the order of 10^{16}, resulting in a severely ill-conditioned problem.

We first consider the Gaussian measurement error case. The image used to generate the data is given by the dashed line in each of the plots of Figure 9.1.

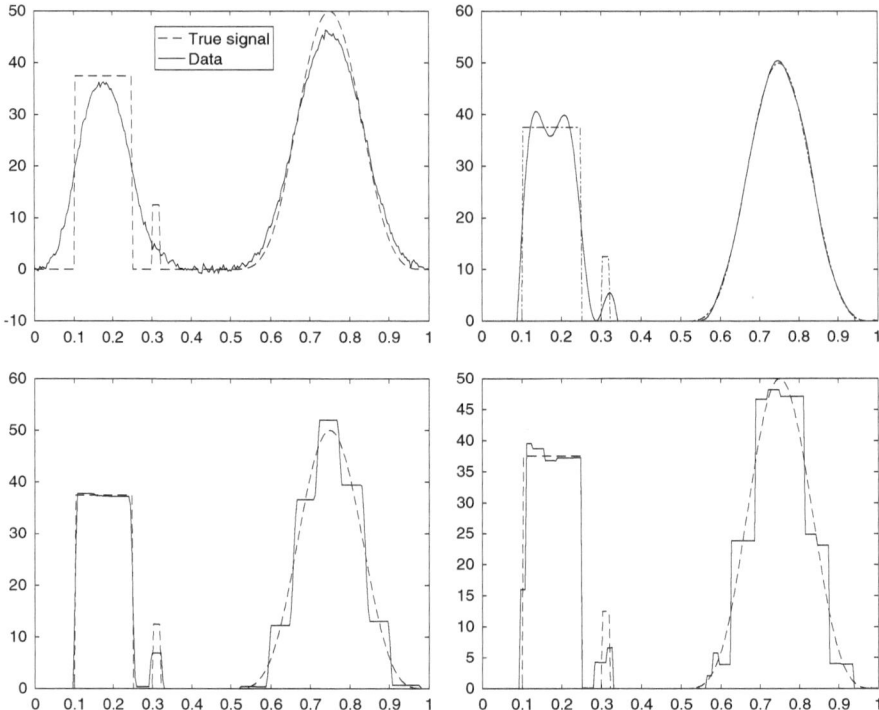

Fig. 9.1 Gaussian measurement error. Upper-left: the true signal and blurred noisy data are given simultaneously. Upper-right: the energy preservation reconstruction. Lower-left: the reconstruction obtained using total variation regularization. Lower-right: the reconstruction obtained using Haar wavelet regularization. All reconstructions are compared to the true signal

The data d is plotted in the upper-left in Figure 9.1 and is generated using (9.1) with the noise variance σ^2 chosen so that the noise strength is 2% that of the signal strength. We estimate the unknown image x first using energy preservation regularization. The result, for one realization of measurement error, is given in the upper-right in Figure 9.1; note that a regularization parameter does not need to be chosen with this approach. The remaining parameters to be defined in Algorithm 9.1 are $\mu_0 = 10^{-2}$ and $\varepsilon = 10^{-8}$. Next, we compute the total variation and wavelet-regularized solutions, both with regularization parameter $\alpha = 10^{-2}$. The results are given in the lower-left in Figure 9.1 for the TV approach, with $\mu_0 = 10^{-2}$ and $\varepsilon = 10^{-2}$, and in the lower-right in Figure 9.1 for the wavelet approach, with $\mu_0 = 10^{-2}$ and $\varepsilon = 10^{-3}$. The SNR is 5.5369, and the relative error of the reconstructions is given in Table 9.1.

Next, we consider the Poisson measurement error case. The image used to generate the data is given by the dashed line in each of the plots of Figure 9.2. The data d is plotted in the upper-left in Figure 9.2 and is generated using (9.2) with the signal intensity chosen so that the noise strength is 4% that of the signal strength.

Table 9.1 The relative error of the reconstructions for the energy preserving, total variation, and Haar wavelet approaches, for both the Gaussian and Poisson noise cases in one dimension

Method	Relative error	
	Gaussian	Poisson
Energy preserving	0.1424	0.1353
Total variation regularization	0.1352	0.1290
Haar wavelet regularization	0.1756	0.1809

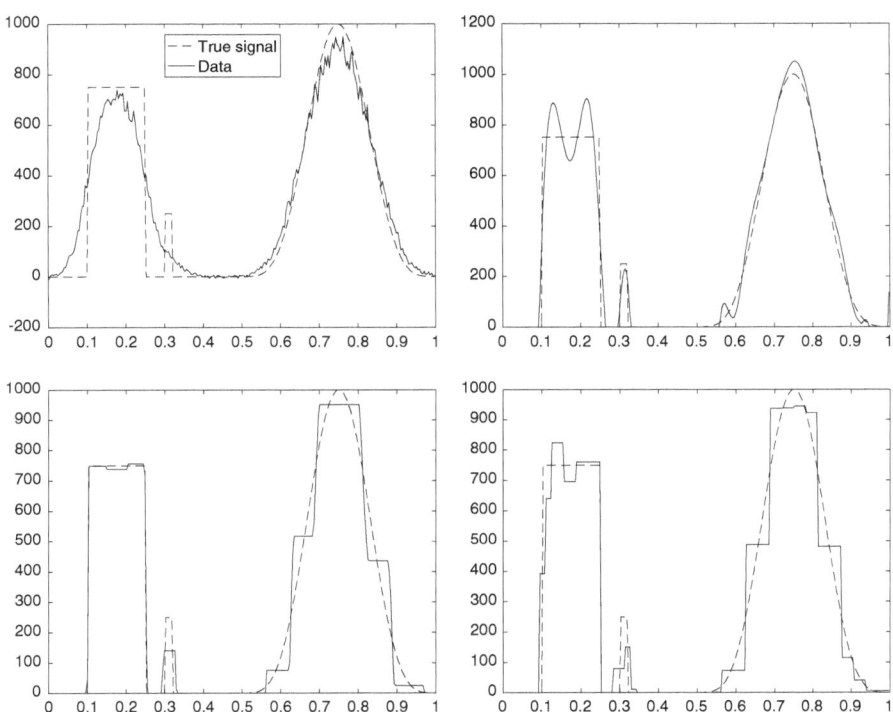

Fig. 9.2 Poisson measurement error. Upper-left: the true signal and blurred noisy data are given simultaneously. Upper-right: the energy preservation reconstruction. Lower-left: the reconstruction obtained using total variation regularization. Lower-right: the reconstruction obtained using Haar wavelet regularization. All reconstructions are compared to the true signal

We estimate the unknown image x first using energy preservation regularization. The result, for one realization of measurement error, is given in the upper-right in Figure 9.2; note that a regularization parameter does not need to be chosen with this approach. The remaining parameters to be defined in Algorithm 9.1 are $\mu_0 = 100$ and $\varepsilon = 10^{-8}$. Next, we compute the total variation and wavelet-regularized solutions, both with regularization parameter $\alpha = 5 \times 10^{-3}$. The results are given in the lower-left in Figure 9.2 for the TV approach, with $\mu_0 = 10^{-2}$ and $\varepsilon = 10^{-6}$, and in the lower-right in Figure 9.2 for the wavelet approach, with

$\mu_0 = 10^{-1}$ and $\varepsilon = 10^{-6}$. The SNR is 24.7209, and the relative error of the reconstructions are given in Table 9.1.

Finally, we note that we have also implemented the ADMM algorithm of [7] on this problem. This method also creates an augmented Lagrangian function but with additional penalty terms added to account for the bound constraints. This has the benefit that the quadratic minimization problem for updating x is unconstrained and hence simpler and more efficient to solve. However, this simplicity comes at a cost. First off, in our experiments, incorporating the nonnegativity constraints directly into the problem, while more computationally demanding, yields better results. Second, the additional terms added to the augmented Lagrangian require significantly more tuning: in [7], additional parameters β_1 and β_2 are needed in order to weight the updates of the corresponding Lagrange multipliers. And finally, in ADMM, the authors fix the penalty parameter μ, whereas we increase μ, allowing a proof of convergence to the solution of the original problem (9.12).

9.4.2   Two-Dimensional Synthetic Image Deconvolution

In many inverse problems applications, e.g., imaging, the spatial domain is two-dimensional. Thus we must show that our method is also effective on two-dimensional problems. Two-dimensional convolution has the form

$$d(s, t) = \int_0^1 \int_0^1 K(s - s', t - t') f(s', t') ds' dt'.$$

As above, we choose a Gaussian convolution kernel K and discretize using midpoint quadrature on an 128×128 uniform computational grid over $[0,1] \times [0,1]$. The image used to generate the data is a standard synthetic satellite from the literature. The blurred, noisy data d is plotted in the upper-left in Figure 9.3 and is generated using (9.1) with the noise variance σ^2 chosen so that the noise strength is 2% that of the signal strength.

We first consider the Gaussian measurement error case. We estimate the unknown image first using the energy preservation approach of (9.6). The result is given in the upper-right in Figure 9.3 with $\mu_0 = .1$ and $\varepsilon = 10^{-5}$. Next, we compute the TV and wavelet-regularized solutions with the following choices of parameters in both cases: $\alpha = 0.01$, $\mu_0 = 10^{-3}$, and $\varepsilon = 10^{-2}$. The results are plotted in the lower-left (TV) and lower-right (wavelet) in Figure 9.3. The relative errors for the three reconstruction methods are given in Table 9.2.

Next, we consider the Poisson measurement error case. The data d is plotted in the upper-left in Figure 9.4 and is generated using (9.2) with the signal intensity chosen so that the noise strength is 5% that of the signal strength. We estimate the unknown image x first using energy preservation regularization. The result, for one realization of measurement error, is given in the upper-right in Figure 9.4; note

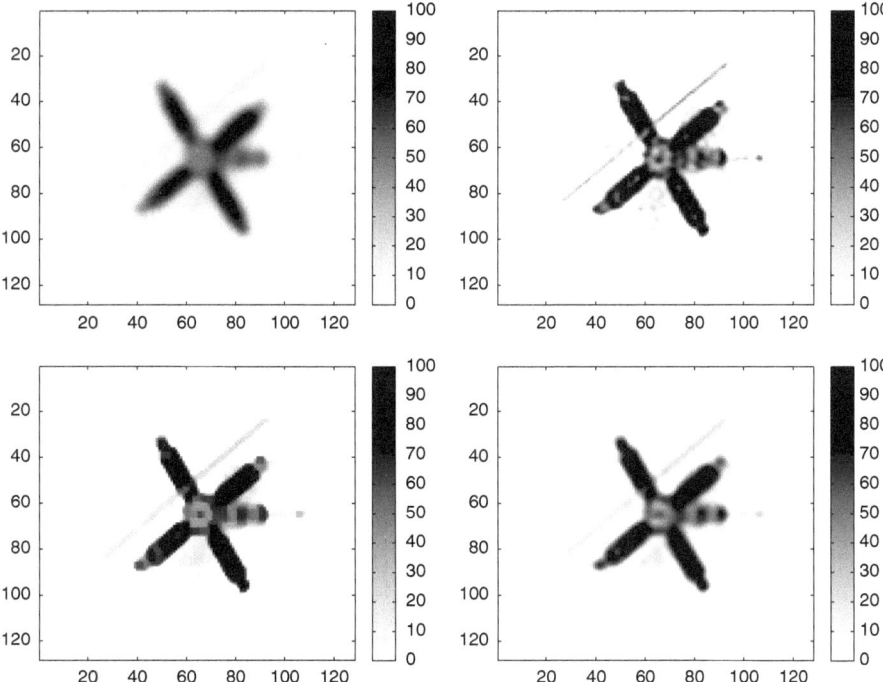

Fig. 9.3 Gaussian measurement error. Upper-left: the blurred noisy data. Upper-right: the energy preservation reconstruction. Lower-left: the reconstruction obtained using total variation regularization. Lower-right: the reconstruction obtained using Haar wavelet regularization

Table 9.2 The relative error of the reconstructions for the energy preserving, total variation, and Haar wavelet approaches, for both the Gaussian and Poisson noise cases in two dimensions

	Relative error	
Method	Gaussian	Poisson
Energy preserving	0.0600	0.1849
Total variation regularization	0.1520	0.1795
Haar wavelet regularization	0.1725	0.1862

that a regularization parameter does not need to be chosen with this approach. The remaining parameters to be defined in Algorithm 9.1 are $\mu_0 = 100$ and $\varepsilon = 10^{-8}$. Next, we compute the TV and wavelet-regularized solutions, with the following choices of parameters, in both cases using the parameter choices: $\alpha = 10^{-3}$, $\mu_0 = 10^{-1}$, and $\varepsilon = 10^{-3}$. The results are plotted in the lower-left (TV) and lower-right (wavelet) in Figure 9.4. The relative errors for the three reconstruction methods are given in Table 9.2.

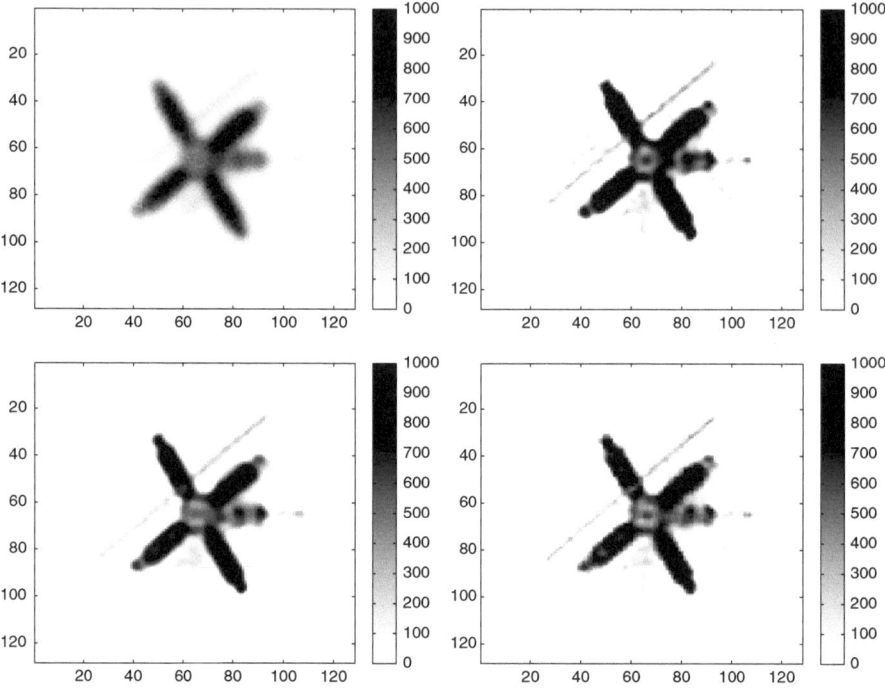

Fig. 9.4 Poisson measurement error. Upper-left: the blurred noisy data. Upper-right: the energy preservation reconstruction. Lower-left: the reconstruction obtained using total variation regularization. Lower-right: the reconstruction obtained using Haar wavelet regularization

9.4.3 One-Dimensional Real Image Deconvolution

In the security sciences, image reconstruction of pulsed-power X-ray radiography is often used to determine image features or the density of the object being imaged [9, 14, 17]. A pulsed-power source generates X-rays that pass through the object(s) to be imaged, and the X-rays not attenuated in the scene are absorbed by a scintillator, a material which luminesces visible light when excited by ionized radiation. The visible light is collected on a CCD array, which produces a radiograph. Calibration objects, such as the step wedge shown in Figure 9.5, are used to compute the X-ray transmission curve for a given radiograph since the X-ray spectrum and intensity vary from shot to shot. The transmission curve, which describes the mapping of object areal density onto image density, can be determined by knowing the areal densities of the calibration object.

The radiograph of the tantalum step wedge in Figure 9.5a was taken from the Cygnus dual-beam radiographic imaging facility at the US Department of Energy's Nevada National Security Site, a 2.25 MeV endpoint energy, pulsed-power X-ray source with a rod-pinch diode [13, 14]. The step wedge object is a uniform

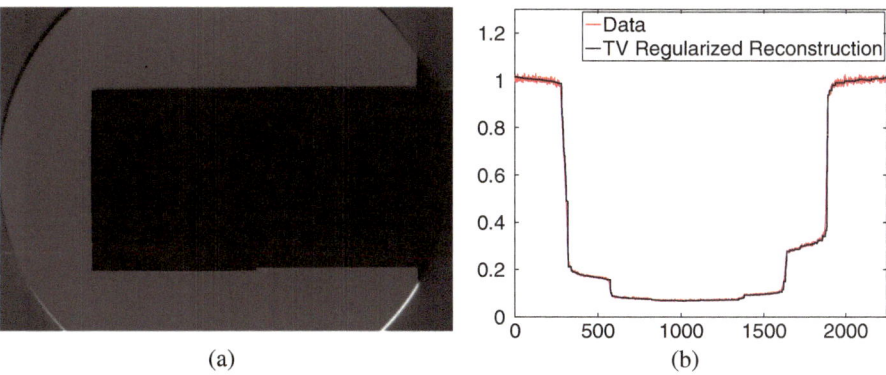

(a)   (b)

Fig. 9.5 (**a**) A radiograph of a so-called step wedge used for object density reconstruction. The goal is to distinguish among the darkest shades of gray. (**b**) A vertical lineout, or column (red), from the left image together with the reconstruction (black)

material made up a discrete number of different thicknesses. The thinner the step, the more X-rays that pass through to be absorbed by the scintillator, producing a greater intensity at the CCD array. The point spread function, or convolution operator, for an X-ray pulse at Cygnus is calculated from a radiograph of an "L" rolled edge [4]. With the lineout (cross-sectional) data and point spread function in hand, we deconvolve the data using the one-dimensional total variation approach with $\alpha = 0.05$. CCD cameras are contaminated by both Gaussian and Poisson noise processes, and here we approach the reconstruction with a Gaussian noise assumption [8]. The reconstruction is plotted together with the data in Figure 9.5b. The remaining parameters to be defined in Algorithm 9.1 are $\mu_0 = .01$, $\mu_{max} = 10$, $\lambda_0 = 1$, $\beta = 3$, and $\varepsilon = 10^{-6}$.

9.5 Conclusion

Ill-posed, linear inverse problems are common throughout the literature. One of the most common examples of such problems is image deblurring. The problem of solving an image deblurring problem is ill-posed (i.e., unstable) without some form of regularization. Here we implement regularization via the addition of constraints to a convex (or convex-quadratic) optimization problem derived from the assumption of either Gaussian or Poisson measurement error. The constraints are chosen to impose certain physical qualities to the unknown image, e.g., nonnegativity, smoothness, and the presence of sharp edges. We focus here on two forms of regularization. The first we call energy preservation, where we simply solve the maximum likelihood optimization problem, assuming either Gaussian or Poisson measurement error, with a nonnegativity constraint together with the simple equality constraint $\|f\|_1 = \|d\|_1$, where f is the unknown, d is the observed data, and the blurring matrix K is assumed to preserve energy, i.e., $K^T 1 = 1$. Secondly, we implement regularization by adding an L^1-norm of the gradient or Haar wavelet coefficients to the negative-log likelihood function, again assuming either Gaussian or Poisson measurement error. In this case, we recast the resulting convex, nonnegatively constrained minimization problem as a bound- and equality-constrained convex minimization problem. To solve the problem, we implement an augmented Lagrangian (AL) algorithm that requires the repeated solution of a large-scale bound-constrained, convex optimization subproblem. For this subproblem, we advocate the use of two highly efficient methods: for the Gaussian (convex-quadratic) case, we use gradient projection conjugate gradient, whereas in the Poisson (convex) case, we use gradient projection-reduced Newton. The accurate solution of the bound-constrained optimization problems obtained using these methods leads to good results. We present numerical experiments on synthetic one- and two-dimensional imaging examples showing that the AL method works well. In addition, a radiograph of a calibration object from a pulsed-power X-ray source at a radiography facility of the US Department of Energy was used to demonstrate the applicability of this approach and the features of its reconstruction.

Acknowledgements This work was done in part by National Security Technologies, LLC, under Contract No. DE-AC52-06NA25946 with the US Department of Energy and supported by the Site Directed Research and Development program. The US government retains, and the publisher, by accepting the article for publication, acknowledges that the US government retains a non-exclusive, paid-up, irrevocable, worldwide license to publish or reproduce the published form of this manuscript, or allow others to do so, for US government purposes. The US Department of Energy will provide public access to these results of federally sponsored research in accordance with the DOE Public Access Plan (http://energy.gov/downloads/doe-public-access-plan). DOE/NV/25946--2476.

References

1. Bardsley, J.M.: An efficient computational method for total variation-penalized poisson likelihood estimation. Inverse Probl. Imaging **2**, 167–185 (2008)
2. Bardsley, J.M.: Laplace-distributed increments, the Laplace prior, and edge-preserving regularization. J. Inverse Ill-Posed Probl. **20**, 271–285 (2012)
3. Bardsley, J.M., Vogel, C.R.: A nonnnegatively constrained convex programming method for image reconstruction. SIAM J. Sci. Comput. **25**, 1326–1343 (2004)
4. Barnea, G.: Penumbral imaging made easy. AIP Rev. Sci. Instrum. **65**, 1949–1953 (1994)
5. Beck, A., Teboulle, M.: Fast gradient-based algorithms for constrained total variation image denoising and deblurring problems. IEEE Trans. Image Process. **18**, 2419–2434 (2009)
6. Bertsekas, D.P.: Nonlinear Programming. Athena Scientific, Belmont (1995)
7. Chan, R.H., Tao, M., Yuan, X.: Constrained total variation deblurring models and fast algorithms based on alternating direction method of multipliers. SIAM J. Imaging Sci. **6**, 680–697 (2013)
8. Howard, M., Luttman, A., Fowler, M.: Sampling-based uncertainty quantification in deconvolution of x-ray radiographs. J. Comput. Appl. Math. **270**, 43–51 (2014)
9. Kelley, T.A., Stupin, D.M.: Radiographic least squares fitting technique accurately measures dimensions and x-ray attenuation. Rev. Prog. Quant. Nondestruct. Eval. **17A**, 371–377 (1998)

10. Moré, J.J., Toraldo, G.: On the solution of large quadratic programming problems with bound constraints. SIAM J. Optim. **1**, 93–113 (1991)
11. Mueller, J., Siltanen, S.: Linear and Nonlinear Inverse Problems with Practical Applications. SIAM, Philadelphia (2012)
12. Nocedal, J., Wright, S.J.: Numerical Optimization, 2nd edn. Springer, New York (2006)
13. Oliver, B.V., Berninger, M., Cooperstein, G., Cordova, S., Crain, D., Droemer, D., Haines, T., Hinshelwood, D., King, N., Lutz, S., Miller, C.L., Molina, I. , Mosher, D., Nelson, D., Ormond, E., Portillo, S., Smith, J., Webb, T., Welch, D.R., Wood, W., Ziska, D.: Characterization of the rod-pinch diode x-ray source on Cygnus. In: IEEE Pulsed Power Conference, pp. 11–16 (2009)
14. Smith, J., Carlson, R., Fulton, R., Chavez, J., Ortega, P., O'Rear, R., Quicksilver, R., Anderson, B., Henderson, D., Mitton, C., Owens, R., Cordova, S., Maenchen, J., Molina, I., Nelson, D., Ormond, E.: Cygnus dual beam radiography source. In: IEEE Pulsed Power Conference, pp. 334–337 (2005)
15. Tikhonov, A.N., Goncharsky, A. , Stepanov, V.V. , Yagola, A.G.: Numerical Methods for the Solution of Ill-Posed Problems. Springer, Berlin (1995)
16. Vogel, C.R.: Computational Methods for Inverse Problems. SIAM, Philadelphia (2002)
17. Whitman, R.L., Hanson, H.M., Mueller, K.A.: Image analysis for dynamic weapons systems. Los Alamos, Report LALP-85-15 (1985)

Chapter 10
Shape Patterns in Digital Fabrication: A Survey on Negative Poisson's Ratio Metamaterials

Bengisu Yılmaz, Venera Adanova, Rüyam Acar, and Sibel Tari

Abstract Poisson's ratio for solid materials is defined as the ratio of the lateral length shrinkage to the longitudinal part extension on a simple tension test. While Poisson's ratio for almost every material in nature is a positive number, materials having negative Poisson's ratio may be engineered. We survey computational works toward design and fabrication of negative Poisson's ratio materials focusing on shape patterns from macro to micro scale. Specifically, we cover folding, knitting, and repeatedly ordering geometric structures, i.e., symmetry. Both pattern design and the numerical aspects of the problem yield various future research possibilities.

10.1   Introduction

Metamaterials are materials that have unusual optical [24], acoustic [12], electromagnetic [21], thermal [16], or mechanical [20] properties because of their structure or constitution. Greek prefix "meta-" (-μετα) meaning "beyond" is used to express this purposely engineered materials. Beyond natural or bulk materials, the organization of their structures makes metamaterials respond very differently under certain circumstances. For example, mechanical metamaterials with negative Poisson's ratio, under longitudinal tension, become thicker perpendicular to the applied force.

B. Yılmaz
Sabanci University, Istanbul, Turkey
e-mail: bengi@sabanciuniv.edu

V. Adanova · S. Tari (✉)
Middle East Technical University, Ankara, Turkey
e-mail: venera@ceng.metu.edu.tr; stari@metu.edu.tr

R. Acar
Okan University, Istanbul, Turkey
e-mail: ruyam.acar@okan.edu.tr

© The Author(s) and the Association for Women in Mathematics 2018   161
A. Genctav et al. (eds.), *Research in Shape Analysis*, Association for Women in Mathematics Series 12, https://doi.org/10.1007/978-3-319-77066-6_10

Recently, due to technology trends toward digital fabrication via printing, a form of additive manufacturing called mechanical metamaterials became quite popular. In this survey, we focus on negative Poisson's ratio mechanical metamaterials.

The macroscale or bulk behavior expressed via negative Poisson's ratio is a result of the purposeful engineering of microstructures. The bulk behavior is an emergent behavior, which is produced by the arrangement of smaller units. Hence, a metamaterial is a tessellation of smaller units with certain shapes; these units serve as the unit cells of a tessellation. Using computational methods microstructures leading to desired bulk behaviors can be designed. It is also possible to engineer the behavior at a macroscale, e.g., by forming a folded surface using origami-like folds. In that case, the underlying unit cells are tessellated folding patterns. A metamaterial is a patterned repetition of shapes.

Our survey covers a wide range of computational methods targeting the design of shapes and patterns that give rise to a desired bulk behavior. We start with the developments in materials science in order to set the scope. Then, we survey computational works that address engineering negative Poisson's ratio metamaterials.

10.2   Negative Poisson's Ratio

Poisson's ratio for solid materials is defined as the ratio of the lateral length shrinkage to the longitudinal part extension on a simple tension test. The relationship between shear modulus (G), bulk modulus (K), and Poisson's ratio (ν) for isotropic linear elastic materials is defined as follows:

$$G := \frac{3K(1 - 2\nu)}{2(1 + \nu)}$$

The most common linear elastic and isotropic materials have Poisson's ratio equal to $1/3$, while Poisson's ratio of rubbery materials may reach almost $1/2$ since the bulk deformation is reasonably hard to happen while the materials goes under shear deformation easily ($G << K$). While for almost every material in nature Poisson's ratio corresponds to a positive number, theory of elasticity [40] with energy arguments limits Poisson's ratio to the interval $[-1, 0.5]$ for isotropic linear elastic materials. That is, there can be materials with negative Poisson's ratio. A negative Poisson's ratio material behaves quite the opposite of rubber, $K << G$, easily deformed on volumetric scale and hardly undergoes shear deformation. This behavior on natural materials is not common although designed structures on some materials show negative Poisson's ratio properties. In other words, materials having negative Poisson's ratio may be observed under certain conditions in nature or by designing intentionally. For instance, two-dimensional conventional honeycomb structures composed of regular hexagons have Poisson's ratio of $+1$ under in-plane stress. On the other hand, by converting two convex sides of hexagon into concave edges, honeycomb cells may result with negative Poisson's ratio such as -1 [20]. These inverted cells, also called reentrant cells, are the key to this behavior called

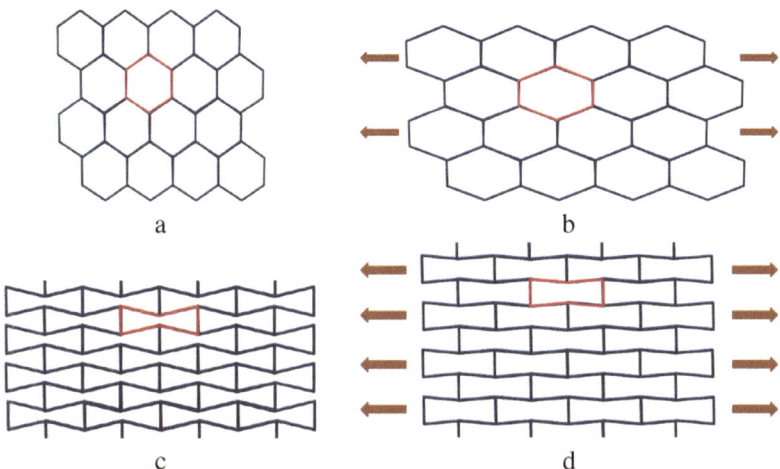

Fig. 10.1 Honeycomb structures: (**a**) conventional honeycomb; (**b**) effect of stretching (a); (**c**) inverted honeycomb, $v = -1$; (**d**) effect of stretching (c)

auxetics (Figure 10.1). In the most generic study of its time, Lakes has showed several types of materials with negative Poisson's ratio [20]. However the name *auxetic* for mechanical metamaterials was not used in the original study. The first study calling the materials with negative Poisson's ratio as *auxetic* is reported by Evans around same time period [8]. Since only two decades has passed from the first study, the research area is still full of open questions. Below, we cover several examples including foams, fibers, and composites.

Polymer foam structures and ductile metallic foams can be designed with inverted or reentrant cell structures resulting in negative Poisson's ratio values [10, 19]. In the foam structures, by the unfolding of the reentrant cells, the negative Poisson's ratio magnitude may be increased and isotropy can be achieved. For anisotropic micro-cellular foams, unintentional negative Poisson's ratios may be observed in some directions. For example, in 1989, a high molecular weight polymer, PTFE (polytetrafluoroethylene), has been processed as sheet and cylindrical ribbon with remarkably large negative Poisson's ratio up to −12. The microstructure of the polymer composed of an open network of anisotropic disk-shaped particles and fibre network with a range of 30% to 80% empty volume. The stress-strain curve inquiry of expended PTFE cylindrical ribbons shows three different mechanical behavior: under low modulus deformation, material undergoes large strain with particle translation and rotations occurring (maximum negative v); as Young's modulus increases, material behaves like elastic spring (reversible deformation/reaching $v = 0$); on continued deformation stage, material goes under plastic deformation with positive Poisson's ratio [4, 9].

In early studies, crystalline and amorphous structures that are reported in micro-nano scale unusually show the negative Poisson's ratio effects under determined conditions. For instance, Poisson's ratio of highly anisotropic single crystalline arsenic can vary in the range of $[-1,+1.5]$ depending on the interest of direction [13]. Crystalline forms of silica exhibits Poisson's ratio from -0.5 to $+0.008$, similarly on the interest of direction inside crystal [14]. The negative Poisson's ratios are also observed for single-crystal cadmium, antimony, and bismuth that have highly anisotropic internal structures [13, 23], for pyrolytic graphite under thermal protection [11], and for certain ferromagnetic films with certain concentration values [29]. Due to the directional anisotropy of internal structures, negative Poisson's ratio can be observed under certain conditions for chemical compounds and elements.

Concerning the macroscale structured metamaterials, the negative Poisson's ratio of fiber-reinforced composite laminates has been reported in certain stacking sequences only on a small range of orientation angle of the applied force [25, 41]. The desired negative Poisson's ratio can be achieved by the intentional change on stacking sequence of composite laminate by conserving isotropy on either two or three dimensions [26].

10.3   Engineering the Auxetic Behavior

We discuss computational works toward designing materials with auxetic behavior in three groups.

10.3.1   Macrostructure

At the macrostructure level, auxetic behavior can be obtained by the so-called meta-surfaces constructed via folding. The fold concept may have several indications, mostly focusing on specific shapes such as creases or pleats in different scales, for example, folding of graphene layers [5, 33], unfolding of tree leaves [17], and building of mountains [21]. The folding concept had become even more significant by the invention of paper which allowed the creation of the folded structures. Folding paper which is called origami in ancient Japan is a combination of concatenating "oru" (fold) and "kami" (paper). Paper, which is thin enough and yet inextensible, allows the construction of folded structures with a variety of shape options. In the last decades, the connection between origami structures and mathematics has even created a new research area called origami mathematics.

Origami mathematics combined with kinematics and structural mechanics leads to innovations, mostly at a theoretical level. Applications are focused in civil engineering, architecture, and design. The study of the folding structure combined with the building methods for architectural components opens a wide range of research questions on the form-finding process. Since origami is continuous and inexpensive, it is a good candidate for small robot design and microelectromechanical systems (MEMS) [21]. There have been many studies regarding the mathematical theories and kinematics of pleats and creases from single-vertex folds to curved fold-forming structures. Also three-dimensional folded tessellations (periodically folded patterns) have been investigated as one of the most attractive areas of this research. The reason for the recent increase in interest is the accessible shapes of the "flat" tessellation. It is remarkable that after microstructuring the original flat surface with a periodic pattern, new geometries become possible [35]. This is the reason why these surfaces are called "meta-surfaces" in reference to metamaterials, both having finely tuned microstructures [32, 43, 45].

During the *meta-surface* tessellation research, the most investigated pattern is Miura-ori with different perspectives. Research studies can be categorized as the homogenized behavior of the Miura pattern, sandwich panel applications, symmetric approaches to a generalized folding family of tessellation, perturbation of patterns to free forms, morphing structures, and active origami shapes [21].

Despite a surge of interest in understanding the mathematics of folded structures and qualitative studies on origami, most work has been limited to two-dimensional behavior – either auxetic behavior in-plane or bending of one-dimensional corrugated strips – until the study of Wei et.al. on basic mechanical properties of origami [43], where they investigated three-dimensional elastic response, Poisson's ratio, and rigidities on in-plane stretching and out-of-plane bending conditions.

An alternative solution to make a flat material (sheet metal, plastic, or leather) more extensible is proposed by Konaković et al. [18]. The problem is that given a thin-plate material which can bend freely and is flexible, it cannot cover the surface completely because it does not stretch and is inextensible. The proposed solution is to insert specific pattern cuts into those thin-plate materials. This adds more freedom to materials making it possible to approximate the surfaces. The specific pattern cuts make inextensible materials auxetic. In [18] there is a focus on triangular patterns which have kinematic linkage composed of equilateral triangles. Those patterns scale roughly isotropic. The method is as follows. Given a target surface, they find a conformal parameterization for it taking the surface into 2D domain. Then the auxetic material is overlaid on top of the conformal map. Using the conformal map, the auxetic material is taken back to the 3D domain and becomes an initialization for the nonlinear optimization solver. At the optimization stage, an objective function is minimized with respect to vertex positions. It pushes the material as close as possible to the target surface while keeping the equilateral triangles rigid and avoiding collisions between the neighboring triangles.

10.3.2 Between Macro- and Microstructure

Between macro- and microstructuring, the study of three-dimensional textiles is another interesting direction. Its auxetic behavior was studied in [42]. Even though fibers [1, 30, 31], yarns [38, 44], and two-dimensional knitted fabric structures with negative Poisson's ratio have been studied previously, there has not been enough work with the three-dimensional auxetic behavior of fabrics until that study, which showed that the auxetic effect of warp-knitted structures can change in stretching directions (highest auxetic effect on weft direction to lowest auxetic effect on knit direction) and the auxetic behavior of the warp-knitted textile maintains about 65% under 10 cycles of extension. Unlike conventional 2D fabrics, 3D fabrics consist of two layers (face layers; each can be of a different fabric) joined by a microfilament yarn (middle layer). In [42] a new geometric structure for the design of 3D fabrics is proposed. Using the proposed geometric structure, four auxetic warp-knitted space fabrics are designed. It should be noted that though many geometric structures give auxetic behavior, they must be amenable to fabrication into a warp-knitted fabric. The proposed geometric structure has a repeating unit that contains of two parallelograms arranged in a V form. Each parallelogram is divided into six ribs. They are only of two types: long ribs and short ribs. Long ribs are of length $l1$ and $l2$ for short ribs. The angle formed between short and long ribs is given as θ. Under the tension the ribs rotate around their connecting points, giving an auxetic effect to the structure. The face layers of exemplar fabrics are fabricated with non-auxetic polyester filaments. Thus the auxetic effect comes only with the geometric structure.

10.3.3 Microstructure

The importance of micro-nanoscale structured materials (MNSM) is that their mechanical behavior is more predictable by numerical simulations since their properties mostly depend on repeatedly ordered geometric structures [22]. However, due to scalability issues during simulations, predicted mechanical responses of these materials (such as deformations and plastic processes) may not be reachable. At macroscale, the only force acting on the mechanical deformation is gravity. On the other hand, as the scale decreases to micro and nano, the internal structures (e.g., crystal domains) and the specific surface area that depends on the periodicity of unit cells and the stimuli (e.g., electromagnetic waves) play significant roles. Another important property of MNSM is that they can be designed to obtain specified structural anisotropy parameters.

Several recent studies in computer graphics address material fabrication based on microscale structuring [3, 15, 28, 34]. In order to overcome the computational issues at the microscopic level, such as stability and efficiency, these studies use various forms of the homogenization method in numerical computations. The idea in using homogenization is to transfer the computations to coarser (mesoscopic or macroscopic) scales while still modeling the microscopic behaviors. Kharevych et al. [15] were the first to propose a numerical method for the simulation of heterogeneous elastic materials based on recent advancements in homogenization theory [27].

Homogenization is a downsampling process on differential equations where the aim is to replace highly oscillating coefficients with smoother coefficients. In other words, it is a form of averaging to obtain macroscopic behaviors from small-scale (microstructural) levels while maintaining the microscopic properties. The main goal is to solve the equations at coarse scales and reduce numerical complexity. There are various approaches to coarsening; for material fields such as elasticity which include rapidly varying fluctuating components, linear (or nonlinear) averaging methods cannot be used. Thus several methods address the issue of homogenizing the material fields without losing the microscopic information. The classical approach proposed by Sigmund et al. [36] uses asymptotic expansion of displacement fields to separate the average and fluctuating parts of elasticity strain. The original theory assumes the periodicity and ergodicity of the material fields. Recently, Ohwadi et al. [27] removed these assumptions by transferring a new metric from small scales to large scales. Their method is based on scalar fields.

Kharevych et al. [15] extended this approach to elasticity tensors also by using a similar metric based on global harmonics. In order to be able to coarsen spatially varying elasticity fields, they use a measurement of the behavior of the fine mesh under linear forces and for given conditions. This measurement of deformations represents a set of displacements called harmonic displacements. Harmonic displacements at the coarse level are then obtained by linear averaging. The key to down sample elasticity without losing fine-scale information is to enforce the potential energy on the coarse grid cells to be equal to the integral of the potential energy over the fine mesh cells by using these displacements in deformation gradient. Coarse elasticity tensor obtained in this way allows efficient simulation of heterogeneous elastic materials with microstructural variations as well as the simulation of anisotropic elastic objects.

Another useful property of the homogenization theory is that it provides a solution to the material design problem through inverse homogenization. Given the coarse fields obtained from homogenization, materials with desired properties can be constructed with optimization modeling. Optimization can be formulated to find the material microstructures with optimum properties (minimum volume, weight, etc.) for the given elasticity properties in which case it corresponds to topology optimization. Another important property of the inverse homogenization process is that it allows the design of negative Poisson's ratio materials.

The works in [3, 28, 34] also address material design with complex heterogeneous elastic behaviors. In [3] Bickel et al. use a form of coarsening based on radial basis functions (RBF). They obtain nonlinear material properties from sampling local, linear deformations of base materials using RBF interpolation. They also use RBF fitting and FEM simulation to obtain the behavior of combinations of material objects. This first part of their model involves material measurement. In the second part, they use an optimization process (inverse homogenization) to obtain the properties of a combination of stacked layers of materials, that is, composite structures, that match the deformation behavior of the target object. Each layer in the composite structure contains one type of base material. Here, all base materials are homogeneous, while the final output is designed from inhomogeneous materials. All base materials exhibit nonlinear hyper-elastic behavior. To solve the optimization problem for each layer of the target object material, the branch-and-bound algorithm (based on a decision tree) is used. For that, given a desired object shape, it is divided into N regular cells. For each cell a single material is chosen from M possible base materials. This choice of a base material is called a *design*. At each level of the tree, possible options for only one cell are considered. Observe that this is an exponential problem yielding slow computation.

Panetta et al. [28] also use the homogenization method to obtain homogenized elasticity tensors. The goal is again to find material structures using these homogenized fields in the optimization. They assume periodicity and use the method based on asymptotic expansion in homogenization of the elasticity tensors. Accordingly, the displacement term is assumed to consist of an average (linear) and a fluctuating part. The elasticity tensor (which is called the base tensor) corresponding to this microscopic displacement is obtained by solving the elasticity equations with a FEM approach. The homogenized elasticity tensor is then obtained by averaging this base tensor. When this homogenized tensor is used in optimization (inverse homogenization) to obtain the material structure in a cell, infinitely tiling the space with this structure would give the homogenized elasticity tensor. Inverse homogenization is used for a similar purpose as in [3] to find the best combination of materials from a library; however here the goal is shape optimization [2]. Unlike topology optimization topological changes are not allowed in shape optimization. This is a limitation in material design; however it simplifies the design process. Once the material properties are obtained, pattern search from the library is performed. The construction of a microstructure library consists of two parts: topology enumeration and shape optimization. In the topology enumeration part, all possible topologies within a unit cell are enumerated. The work is focused on cubic structures. Since cube is reflectional symmetric, the symmetry planes split a cube into 48 identical tetrahedrons. This allows to enumerate topologies only for a single tetrahedron and the whole structure is retrieved from reflections. For topology enumeration 15 node samples of representative tetrahedrons are used: 4 vertex nodes, 6 edge nodes, 4 face nodes, and 1 internal node. By changing the connectivity between these 15 nodes, different topological configurations are obtained. After eliminating the topologies that do not meet certain constraints, they obtained 1205 topologies and 138 tileable families. Then, for each topology, a shape optimization is performed with vertex positions and edge thicknesses as parameters, in order to determine the range of achievable material properties. At this stage, the microstructure is fit to an elasticity tensor.

In contrast to [3], the approach in [28] constructs the library of microstructures from a single material. The aim is to design as many microstructures as possible so that a wide range of isotropic (E, v) material space is covered. Then the elasticity properties of a given object are obtained by tiling the microstructures with closest (E, v) pairs for the object with given elasticity. In [3] the library contains the models of specific base materials which are considered to be known. Different elasticity properties are obtained by combining these base materials in terms of composites as mentioned previously.

Schumacher et al. [34] use the coarsening approach of Kharevych et al. [15] to simulate spatially varying elasticity fields and without being limited by the periodicity assumption. They use these coarsened elasticity fields in inverse homogenization for topology optimization. They also use harmonic displacements as in [15]; however they use a different formulation [46] for the optimization problem.

The main objective of this work is also to obtain material structures from a family of structures through optimization. Given an object with specified parameters, for each cell of the object, one structure from a metamaterial family (a collection of microstructures) is computed. The generation of metamaterial family starts with a single microstructure which is obtained via microstructure optimization. Then using dilation and erosion different versions of the initial microstructure is obtained. A single metamaterial family covers a small range of material properties. This leads to the generation of more metamaterial families. Unlike previous methods they formulate the optimization based on a function of the coarse elasticity tensor instead of the elasticity tensor itself. This function represents the choice of the material model from a metamaterial family. Selection is based on how well the microstructure describes the cell properties and how similar it is to the neighboring structures. The latter is dictated by the tileability constraint. Note that with this function, the structures are chosen on a coarse grid. Then microstructures in between are obtained through interpolation. This interpolation is realized as a weighted average of the microstructures with similar elastic properties, where the weights are computed from the inverse distance between the input parameters and metamaterial sample parameters. Normally obtaining structures for spatially varying elasticity fields would require several iterations of the optimization. However, this approach based on the selection function provides a much more efficient computation compared to previous methods.

It should be mentioned that the synthesis level of three approaches [3, 28, 34] significantly differ. Both [3, 34] compute new structures at the synthesis level, while [28] use existing library of microstructures. In [34] each cell of the volume mesh representing the object is assigned with a corresponding microstructure from a look-up table either directly or via material optimization.

10.4 Conclusion and Future Work

We have reviewed recent developments in design and fabrication of materials with auxetic behavior. Both pattern design and the numerical aspects of the problem yield various future research possibilities. Currently all the work discussed in this paper use explicit, fixed mesh methods for the microstructure optimization problem. These so-called density methods [6] obtain shape from an interpolated density distribution as discussed previously. This is the most widely studied numerical solution in topology optimization; the main problem here is to accurately obtain the shape from the smoothed density fields. In a different approach, boundary variation methods [6] use implicit functions such as the level-set and phase-field equations to represent the shape. This relatively more recent approach has not been explored in graphics to our knowledge. This approach allows more accurate and high resolution shape/topology representation since shape and topology are derived directly from tracking the boundary contour. Phase-field method does not require distance calculations and boundary reinitialization as the level-set method, and it allows all topology changes.

We would like to investigate the use of phase-field methods in topology optimization in graphics to obtain high resolution and topologically more complex models with improved numerical stability.

Another interesting line of research to pursue would be origami-inspired material design. Aside from the space efficiency it provides, structure design from folded sheets has several important properties. First, it allows the construction of fold patterns with unconventional geometric structures beyond Euclid's rules [39]. Second and more importantly, due to their flexibility in changing the fold patterns, they allow programmable metamaterial design. This property shows that origami-based material design holds very promising future prospects and possible applications in different fields [37, 39]. As such, it has been a highly active research area recently [7, 37]. We would like to focus mainly on two different problems in this area. First, it would be interesting to study the design of fold structures to obtain certain physical properties [43]. Next, we would like to explore the solutions for the inverse problem of obtaining fold structures that represent a given shape, that is, shape design with origami fold structures.

Acknowledgements The work is funded by TUBITAK via 114E204.

References

1. Alderson, K.L., Alderson, A., Smart, G., Simkins, V.R., Davies, P.J.: Auxetic polypropylene fibres: part 1 - manufacture and characterisation. Plast. Rubber Compos. **31**(8), 344–349 (2002)
2. Allaire, G.: Shape Optimization by the Homogenization Method. Springer, New York (2002)
3. Bickel, B., Bächer, M., Otaduy, M.A., Lee, H.R., Pfister, H., Gross, M., Matusik, W.: Design and fabrication of materials with desired deformation behavior. ACM Trans. Graph. **29**(4), 63:1–63:10 (2010)
4. Caddock, B.D., Evans, K.E.: Microporous materials with negative poisson's ratios. I. Microstructure and mechanical properties. J. Phys. D. Appl. Phys. **22**(12), 1877 (1989)
5. Cranford, S., Sen, D., Buehler, M.J.: Meso-origami: folding multilayer graphene sheets. Appl. Phys. Lett. **95**(12) (2009)
6. Deaton, J.D., Grandhi, R.V.: A survey of structural and multidisciplinary continuum topology optimization: post 2000. Struct. Multidiscip. Optim. **49**(1), 1–38 (2014)
7. Dudte, L.H., Vouga, E., Tachi, T., Mahadevan, L.: Programming curvature using origami tessellations. Nat. Mater. **15**, 583–588 (2016)
8. Evans, K.E.: Auxetic polymers: a new range of materials. Endeavour **15**(4), 170–174 (1991)
9. Evans, K.E., Caddock, B.D.: Microporous materials with negative poisson's ratios. II. Mechanisms and interpretation. J. Phys. D. Appl. Phys. **22**(12), 1883 (1989)
10. Friis, E.A., Lakes, R.S., Park, J.B.: Negative poisson ratio polymeric and metallic foams. J. Mater. Sci. **23**(12), 4406–4414 (1988)
11. Garber, A.M.: Prolytic materials for thermal protection systems. Aerosp. Eng. **22**, 126–137 (1963)
12. Gibson, L.J., Ashby, M.F.: Cellular Solids. Cambridge University Press, Cambridge (1997)
13. Gunton, D.J., Saunders, G.A.: The young's modulus and poisson's ratio of arsenic, antimony and bismuth. J. Mater. Sci. **7**(9), 1061–1068 (1972)
14. Haeri, A.Y., Widner, D.J., Parise, J.B.: Elasticity of α-cristobaliite: a silicon dioxide with negative poisson's ratio. Science **257**, 650–652 (1992)

15. Kharevych, L., Mullen, P., Owhadi, H., Desbrun, M.: Numerical coarsening of inhomogeneous elastic materials. ACM Trans. Graph. **28**(3), 51:1–51:8 (2009)
16. Kim, W., Zide, J., Gossard, A., Klenov, D., Stemmer, S., Shakouri, A., Majumdar, A.: Thermal conductivity reduction and thermoelectric figure of merit increase by embedding nanoparticles in crystalline semiconductors. Phys. Rev. Lett. **96**(4), 045901–045904 (2006)
17. Kobayashi, H., Kresling, B., Vincent, J.F.V.: The geometry of unfolding tree leaves. Proc. R. Soc. Lond. B Biol. Sci. **265**(1391), 147–154 (1998)
18. Konaković, M., Crane, K., Deng, B., Bouaziz, S., Piker, D., Pauly, M.: Beyond developable: computational design and fabrication with auxetic materials. ACM Trans. Graph. **35**(4), 89: 1–89:11 (2016)
19. Lakes, R.: Foam structures with a negative poisson's ratio. Science **235**(4792), 1038–1040 (1987)
20. Lakes, R.: Advances in negative poisson's ratio materials. Adv. Mater. **5**(4), 293–296 (1993)
21. Lebee, A.: From folds to structures, a review. Int. J. Space Struct. **30**(2), 55–74 (2015)
22. Lee, J.H., Singer, J.P., Thomas, E.L.: Micro-/nanostructured mechanical metamaterials. Adv. Mater. **24**(36), 4782–4810 (2012)
23. Li, Y.: The anisotropic behavior of poisson's ratio, young's modulus, and shear modulus in hexagonal materials. Phys. Status Solidi (a) **38**(1), 171–175 (1976)
24. Maldovan, M., Thomas, E.L.: Periodic structures and interference lithography. In: Periodic Materials and Interference Lithography: For Photonics, Phononics and Mechanics. Wiley-VCH Verlag GmbH and Co. KGaA, Weinheim (2008)
25. Miki, M., Murotsu, Y.: The peculiar behavior of the poisson's ratio of laminated fibrous composites. Trans. Jpn. Soci.Mech. Eng. Ser. A **54**(501), 970–976 (1988)
26. Milton, G.: Composite materials with poisson's ratios close to — 1. J. Mech. Phys. Solids **40**, 1105–1137 (1992)
27. Owhadi, H., Zhang, L.: Metric-based upscaling. Commun. Pure Appl. Math. **60**(5), 675–723 (2007)
28. Panetta, J., Zhou, Q., Malomo, L., Pietroni, N., Cignoni, P., Zorin, D.: Elastic textures for additive fabrication. ACM Trans. Graph. **34**(4), 135:1–135:12 (2015)
29. Popereka, M.Y., Balagurov, V.G.: Ferromagnetic films having a negative poisson's ratio. Sov. Phys. Solid State **11**, 2938–2943 (1970)
30. Ravirala, N., Alderson, A., Alderson, K.L., Davies, P.J.: Expanding the range of auxetic polymeric products using a novel melt-spinning route. Phys. Status Solidi B **242**(3), 653–664 (2005)
31. Ravirala, N., Alderson, K.L., Davies, P.J., Simkins, V.R., Alderson, A.: Negative poisson's ratio polyester fibers. Text. Res. J. **76**(7), 540–546 (2006)
32. Schenk, M.: Geometry of miura-folded metamaterials. Proc. Natl. Acad. Sci. U. S. A. **370**(1965), 3276–3281 (2013)
33. Schniepp, H.C., Kudin, K.N., Li, J., Prud'homme, R.K., Car, R., Saville, D.A., Aksay, I.A.: Bending properties of single functionalized graphene sheets probed by atomic force microscopy. ACS Nano **2**(12), 2577–2584 (2008)
34. Schumacher, C., Bickel, B., Rys, J., Marschner, S., Daraio, C., Gross, M.: Microstructures to control elasticity in 3D printing. ACM Trans. Graph. **34**(4), 136:1–136:13 (2015)
35. Seffen, K.A.: Compliant shell mechanisms. Philos. Trans. R. Soc. Lond. A Math. Phys. Eng. Sci. **370**(1965), 2010–2026 (2012)
36. Sigmund, O.: Design of material structures using topology optimization. PhD thesis, Technical University of Denmark (1994)
37. Silverberg, J.L., Evans, A.A., McLeod, L., Hayward, R.C., Hull, T., Santangelo, C.D., Cohen, I.: Using origami design principles to fold reprogrammable mechanical metamaterials. Science **345**(6197), 647–650 (2014)
38. Sloan, M.R., Wright, J.R., Evans, K.E.: The helical auxetic yarn – a novel structure for composites and textiles; geometry, manufacture and mechanical properties. Mech. Mater. **43**(9), 476–486 (2011)
39. Stewart, I.: Mathematics: some assembly needed. Nature **448**(7152), 419 (2007)

40. Timoshenko, S., Goodier, J.N.: Theory of Elasticity. McGraw-Hill, New York, NY (1969)
41. Tsai, S.W., Hahn, H.T.: Introduction to Composite Materials. Technomic, Lancaster, PA (1980)
42. Wang, Z., Hu, H.: 3D auxetic warp-knitted spacer fabrics. Phys. Status Solidi B **251**(2), 281–288 (2014)
43. Wei, Z.Y., Guo, Z.V., Dudte, L., Liang, H.Y., Mahadevan, L.: Geometric mechanics of periodic pleated origami. Phys. Rev. Lett. **110**, 215501 (2013)
44. Wright, J.R., Sloan, M.R., Evans, K.E.: Tensile properties of helical auxetic structures: a numerical study. J. Appl. Phys. **108**(4), 044905 (2010)
45. You, Z.: Folding structures out of flat materials. Science **345**(6197), 623–624 (2014)
46. Zhou, S., Li, Q.: Design of graded two-phase microstructures for tailored elasticity gradients. J. Mater. Sci. **43**(15), 51–57 (2008)